Sustainable Animal Agriculture

Sustainable Animal Agriculture

Editor

Vivek Karwande

scitus
academics

Sustainable Animal Agriculture

Edited by **Vivek Karwande**

ISBN: 978-1-68117-099-2

Library of Congress Control Number: 2015935499

SCITUS Academics LLC,
www.scitusacademics.com
Box No. 4766, 616 Corporate Way,
Suite 2, Valley Cottage,
NY 10989

Notice

Printed in United States of America on Acid Free Paper ∞

Contents

Preface

The agricultural activities are often based on individual producer´s decisions and on their attitudes, knowledge and level of technology. It is however also based on political and economic considerations, attitudes and opinions from the society. Thus, continuously updated scientifically based knowledge, both from an environmental, social and economic view, need to be disseminated and applied with a much increased ambition. Technological facts may be well known, but still strong social and economic reasons and pressure from outside to make short term profits hinders the appropriate application of relevant measures. This is the reason why we have all parts of the sustainability concept covered in our texts: the ecological, the social, the economical, and the institutional/juridical. "Sustainable agriculture" has become a popular way of expressing that what society wants is an environmentally sound, productive, economically viable, and socially desirable agriculture. However, the concept of agricultural sustainability does not lend itself to precise definition. Agriculture is practiced in so many climates and in different cultural contexts, so "sustainable agriculture" cannot possibly imply a special way of thinking or of using farming practices. Sustainable agriculture is an approach to securing the necessary resources for safeguarding global food production, biodiversity reserves, recreation needs, water quality and well developed rural areas and wildlife areas. It can also be an effective means of poverty reduction and of achieving the Millennium Development Goals, as well as means of mitigating climate change. It is also about health, welfare, respect and ethics regarding animals and man, as well as quality of food and feed. In other words they are truly transdisciplinary and represent a new holistic outlook on ecosystem health and sustainable agriculture.

Editor

1

Integrated Strategy for Sustainable Cattle Fever Tick Eradication in USA is Required to Mitigate the Impact of Global Change

Adalberto A. Pérez de León[1], Pete D. Teel[2], Allan N. Auclair[3], Matthew T. Messenger[4], Felix D. Guerrero[1], Greta Schuster[5], and Robert J. Miller[6]

[1]Knipling-Bushland U.S. Livestock Insects Research Laboratory, United States Department of Agriculture – Agricultural Research Service, Kerrville, TX, USA

[2]Department of Entomology, Texas A&M University, College Station, TX, USA

[3]Risk Analysis Systems, United States Department of Agriculture – Animal and Plant Health Inspection Service, Riverdale, MD, USA

[4]Cattle Fever Tick Eradication Program, United States Department of Agriculture – Animal and Plant Health Inspection Service, Veterinary Services, Riverdale, MD, USA

[5]Department of Agriculture, Agribusiness and Environmental Sciences, Texas A&M University-Kingsville, Kingsville, TX, USA

[6]Cattle Fever Tick Research Laboratory, United States Department of Agriculture – Agricultural Research Service, Edinburg, TX, USA

ABSTRACT

The ticks *Rhipicephalus* (*Boophilus*) *annulatus* and *R.* (*B.*) *microplus*, commonly known as cattle and southern cattle tick, respectively, impede the development and sustainability of livestock industries throughout tropical and other world regions. They affect animal productivity and wellbeing directly through their obligate blood-feeding habit and indirectly by serving as vectors of the infectious agents causing bovine babesiosis and anaplasmosis. The monumental scientific discovery of certain arthropod species as vectors of infectious agents is associated with the history of research on bovine babesiosis and *R. annulatus*. Together, *R. microplus* and *R. annulatus* are referred to as cattle fever ticks (CFT). Bovine babesiosis became a regulated foreign animal disease in the United States of America (U.S.) through efforts of the Cattle Fever Tick Eradication Program (CFTEP) established in 1906. The U.S. was declared free of CFT in 1943, with the exception of a permanent quarantine zone in south Texas along the border with Mexico. This achievement contributed greatly to the development and productivity of animal agriculture in the U.S. The permanent quarantine zone buffers CFT incursions from Mexico where both ticks and babesiosis are endemic. Until recently, the elimination of CFT outbreaks relied solely on the use of coumaphos, an organophosphate acaricide, in dipping vats or as a spray to treat livestock, or the vacation of pastures. However, ecological, societal, and economical changes are shifting the paradigm of systematically treating livestock to eradicate CFT. Keeping the U.S. CFT-free is a critical animal health issue affecting the economic stability of livestock and wildlife enterprises. Here, we describe vulnerabilities associated with global change forces challenging the CFTEP. The concept of integrated CFT eradication is discussed in reference to global change.

INTRODUCTION

Global change makes human and animal populations vulnerable to emerging and re-emerging tick-borne diseases. Environmental and climatic changes ascribed to human activity are collectively referred to as global change (Sutherst, 2001; Camill, 2010). The accelerated rate of global change alters the epidemiology of tick-borne diseases. Tick-borne diseases are complex systems subject to shifts in ecological processes influencing tick biology and consequently the epidemiology of pathogens transmitted by ticks (Randolph, 2010; Reisen, 2010; Tabachnick, 2010; Olson and Patz, 2011). This review explores the ramifications of global change on the Cattle Fever Tick Eradication Program (CFTEP) of the United States of America (U.S.) and how these challenges threaten the ability to keep the national cattle herd free of bovine babesiosis by suppressing the invasive cattle fever tick (CFT) vectors, *Rhipicephalus* (*Boophilus*) *annulatus* and *R.* (*B.*) *microplus*. The reader is referred to the review by Aubry and Geale (2011) for details on bovine anaplasmosis, which is another economically important infectious disease of cattle caused by *Anaplasma marginale* and transmitted by CFT.

Research on bovine babesiosis and CFT involves a notable discovery in the history of medical science. Elegant studies by Smith and Kilbourne (1893) revealed that the apicomplexan protozoon *Babesia bigemina* transmitted by *R. annulatus* caused bovine babesiosis in susceptible cattle. We now know that *R. microplus* also transmits *B. bovis*, which causes a more virulent form of bovine babesiosis. Smith and Kilborne's monumental work was soon followed by others reporting the role of various blood-feeding arthropod species, such as lice and mosquitoes, transmitting pathogens to humans and animals causing severe maladies like malaria, yellow fever, typhus, and dengue (Schultz, 2008). Cooper Curtice also made invaluable contributions to the history of CFT and bovine babesiosis research. His detailed investigations describing the life history of CFT laid the foundation for the CFTEP (Logue, 1995). The interaction between Smith, a physician, and Kilborne, a veterinarian, is a fine example of what we now term the One Health (Bulloch, 1935). The One Health concept defines the collaborative effort of multiple disciplines to attain optimal health for people, animals, and our environment (Welburn, 2011). Such approach identified knowledge gaps in human and bovine babesioses and the benefits of its application to the CFTEP are reflected in progress to translate research into tools that can be integrated to minimize the impact of CFT outbreaks on U.S. agriculture, which are described here (Pérez de León et al., 2010).

Bovine babesiosis is a devastating tick-borne disease of cattle that was eradicated from the U.S. by eliminating CFT populations. This significant event in the history of disease eradication campaigns was accomplished through the efforts of the CFTEP managed by state and federal agencies in collaboration with livestock producers from 1907 until 1943. A permanent quarantine zone established in Texas extending over 800 km from Del Rio to Brownsville and delimited in the south by the Rio Grande, known in Mexico as Río Bravo, is managed by the U.S. Department of Agriculture-Animal Health Inspection Service (USDA-APHIS) and the Texas Animal Health Commission (TAHC) to buffer tick and stray animal incursions from Mexico where CFT and bovine babesiosis are endemic (Figure 1A).

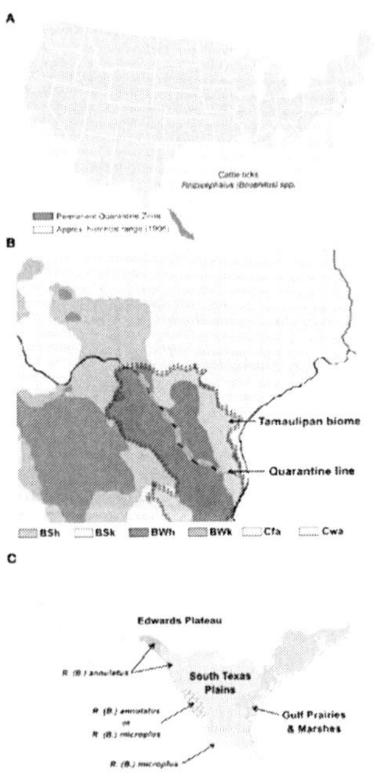

Figure 1: (A,C) The U.S. distribution of cattle ticks in 1906 depicts the zone of risk for tick re-establishment and the permanent quarantine zone along the Texas–Mexico international boundary and quarantine line (A), as adapted

fromGraham and Hourrigan (1977); the Tamaulipan biome of the Texas–Mexico border area (Scifres, 1985), indicated by the overlay with dotted boundary (B), is the major ecoregion comprising the quarantine zone that consists of four Koppen–Geiger climate classifications, as adapted from Peel et al., 2007, where the majority of the biome is of an arid-steppe (BSh) classification linked with the Temperate (Cfa) classification that extends throughout the historical range of cattle fever ticks, whereas in Mexico the biome includes small areas with climate generally classified as Arid desert (BWh) or Temperate (Cwa); the distribution of ecoregions, as adapted from Correll and Johnston (1970), reflects the diversity of ecological and environmental resources (C), and is integrated with an overlay showing the quarantine zone distribution of recorded incursions of *R. annulatus*and *R. microplus*, as adapted from Lohmeyer et al. (2011).

Keeping the U.S. cattle herd free of bovine babesiosis and CFT is an important animal health issue. Previous estimates converted to today's currency rate indicate that the livestock industry realizes annual savings of at least 3 billion dollars since the U.S. was declared free of bovine babesiosis and CFT (Graham and Hourrigan, 1977). The dynamics of global change have increased the risk for re-emergence of CFT and bovine babesiosis in the U.S. (Bram et al., 2002; Guerrero et al., 2007; George et al., 2008). This risk poses significant challenges to future efforts by the CFTEP. Portions of the U.S. still provide suitable habitat for CFT (Estrada-Peña et al., 2005), which correspond with the CFT historical range before the CFTEP was initiated particularly in the 13 states comprising the Southern Region (Figure 1A). Cattle and calves rank among the top five agricultural commodities in nine of the 13 Southern Region states with cash receipts ranging from 5.3% (Florida) to 46% (Oklahoma) of all agricultural commodities. The cow-calf inventory of the 13 Southern Region states is more than 42% of the entire U.S. beef cattle inventory (McBride and Mathews, 2011). Approximately 1/3 of all U.S. fed-cattle originate from Southern Region ranches operated by over 400,000 producers representing nearly 49% of all U.S. cow-calf producers. The economic stability of the cattle industry is vital to the nation.

Global change involves disruptive processes that augment instability in ecosystems, and the services they provide, that are or can be inhabited by ticks (Hanson et al., 2008). These processes add uncertainty to future outcomes of current policies to manage ticks and tick-borne diseases. The disruptive processes driving emergence and re-emergence of tick-borne diseases affecting livestock linked to global change include the alteration of ecosystems and agroecosystems, incursion of natural habitats by humans and animals, shifts in land-use, movement of people and animals, climate change, selection of acaricide–resistant tick populations, increased international trade and travel, civil unrest, expansion of tick and pathogen host range, and governmental or management failure (Pegram et al., 2000; Harrus and Baneth, 2005; George, 2008; De Meeûs et al., 2010; Brouqui, 2011; Perry et al., 2011; Food and Agriculture Organization of the United Nations (FAO), 2011). The deterioration of public safety on the Mexican side of the border in south Texas related to illegal drug activity has complicated efforts by the CFTEP (Aguilar, 2011). Such effects are compounded by the zoonotic nature of some tick-borne diseases and the vulnerability of public health and central veterinary authorities trying to deal with the unexpected consequences of global change (Hoberg et al., 2008;Jones et al., 2008; Black and Nunn, 2009; Munderloh and Kurtti, 2011). Tick species other than *R.microplus* and *R. annulatus* serve as vectors of zoonotic babesioses (Vannier et al., 2008).

Our ability to assess immediate and future impacts of global change on animal pests and diseases needs to be enhanced. Bovine babesiosis and CFT are among the diseases and pests requiring attention (Cumming and Van Vuuren, 2006; Food and Agriculture Organization of the United Nations (FAO), 2008). Agri-intelligence and foresight are two provocative methodologies that have been applied to better understand and proactively recognize threats and risks to national security associated with agriculture to ensure today's decisions anticipate and incorporate adaptive strategies related to an uncertain future (Munroe, 2007; Munroe and Willis, 2007; Willis, 2007; Suk et al., 2008; Pretty et al., 2010). The incorporation of inputs from diverse disciplines and organizations with broad mandates is a goal of agri-intelligence, which enables analysis through networks aimed at synergizing efforts to deliver a value-added output decision makers can use to make more informed policy development and superior operational outcomes (Munroe, 2007). Foresight is a process that looks at the time horizon to facilitate forward preparedness efforts through critical thinking focused on exploring scenarios including those that may appear unlikely today (Smith, 2007). Forecasting approaches, including modeling, are advocated as a means to facilitate anticipation and enhance prevention, preparedness, and the management of disease outbreaks (Baylis and Githeko, 2006; Garner et al., 2007; Munroe and Willis, 2007; Estrada-Peña, 2008; Woolhouse, 2011).

The principles of agri-intelligence and foresight are applied here to dissect the forces of global change and discern the interconnectedness between environmental and anthropogenic factors driving the temporal and spatial variation of CFT outbreaks in south Texas. This exercise was framed by the historical record of CFT outbreaks in the U.S. since 1959 (Figure 2), and it provides further assessment of CFT and bovine babesiosis research needs identified through a process that applied the One Health concept (Pérez de León et al., 2010). The historical situation with CFT is outlined to reveal critical challenges the CFTEP is facing. Such aspects unify the past, present, and future. This perspective provided a lens to recognize and appreciate the complexity of, and the relationships between issues the CFTEP must contend with to contain CFT within the permanent quarantine zone. The correlates and trends with global change described below formed the basis to suggest a framework for the integration of CFT eradication strategies that could be used to enhance the ability of regulatory agencies in the U.S. to anticipate, prepare for, mitigate, and prevent the burden of CFT outbreaks thereby contributing to the benefits society derives from having a sustainable livestock production system (Figure 3). Sustainability in this context means continued productivity into the indefinite future through the use of sound scientific evidence to inform decision making and guide policy makers involved with the CFTEP while maximizing agroeconomic opportunities for livestock producers (National Research Council of the National Academies, 2010; Simmons, 2011; Walker, 2011).

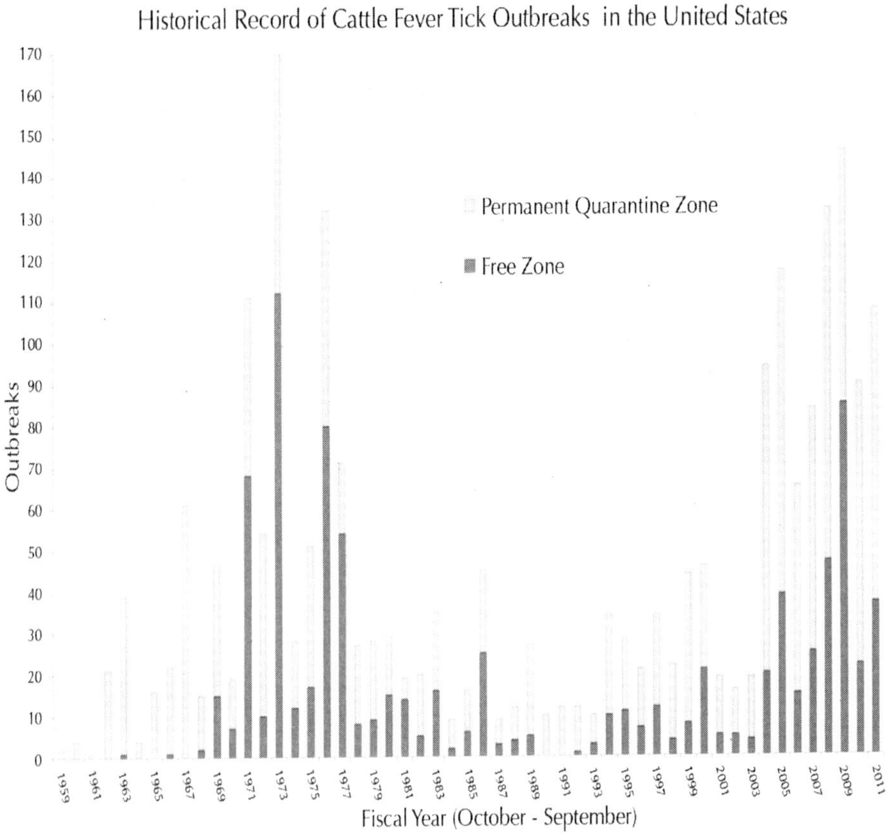

Figure 2: Historical record of CFT outbreaks in the U.S. Each fiscal year shows the contribution by the Permanent Quarantine and Free Zones to the total number of outbreaks. The Free Zone comprises the area in the 48 contiguous states outside the Permanent Quarantine Zone, which is located along the Rio Grande in southern Texas. The period shown in the graph depicts records maintained by USDA-APHIS-Veterinary Services and it covers fiscal years 1959 through 2011. Data updated after Pérez de León et al. (2010).

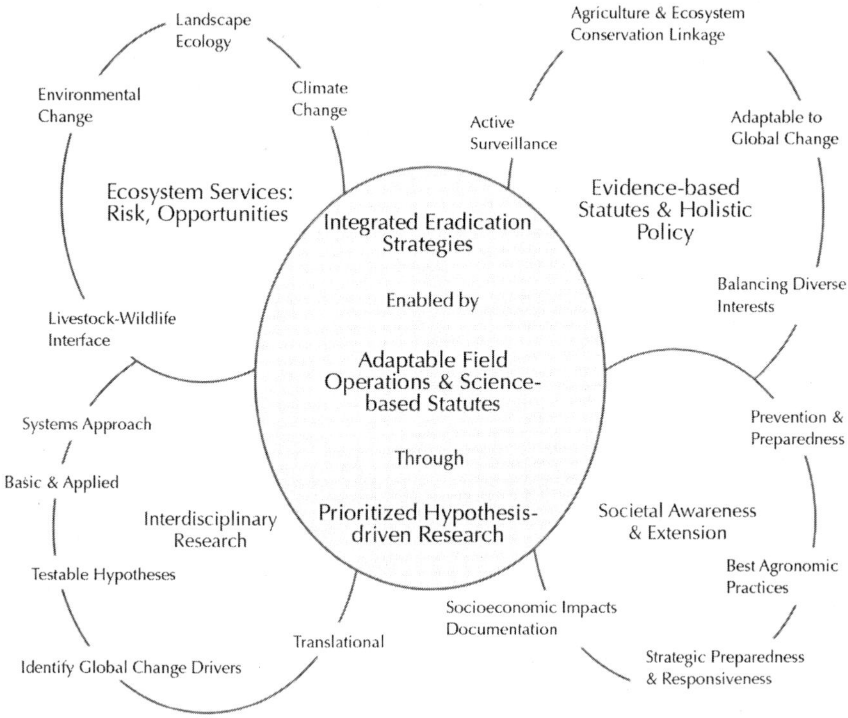

Figure 3: Suggested framework toward integrated cattle fever tick eradication strategies depicting interconnectedness of critical elements for sustainability of bovine babesiosis-free status in the U.S.

ARRIVAL AND SPREAD OF CATTLE FEVER TICKS IN THE U.S

The CFT are not native to the Americas. The origin of *R. annulatus* includes the arid and temperate climates of southern Russia, the Near- and Middle-East, and the Mediterranean basin whereas the origin of *R. microplus* is the tropical climates of the Indian subcontinent (George, 1990; Estrada-Peña et al., 2006). Evidence indicates that Spanish colonialists served as the pathway that brought livestock infested with CFT and infected with bovine babesiosis to Mexico and Cuba during the 1500s (García-García et al., 1999). The introduction and spread of CFT into what is now the U.S. followed the pattern of North American settlement and commerce. It is suspected that by the end of the eighteenth century bovine babesiosis and *R. annulatus* had emerged in the U.S. because in 1795 North Carolina prohibited driving cattle into the state from South Carolina or Georgia between 1 April and 1 November of each year due to an undetermined disease of livestock (George, 1989). By the late 1800's, CFT were distributed throughout the southern states and southern California (Figure1A). Bishopp (1913) first reported cattle infestations with *R. microplus* after collecting the ticks in 1912 from animals in Key West, Florida; he speculated that this CFT arrived from the Caribbean islands through commerce. The ticks and bovine babesiosis, known earlier as Texas, Mexican, Spanish, or splenetic fever, were significant economic impediments to cattle industry development throughout the Southern U.S. (Malone, 1989; Strom, 2009). High mortality rates in non-immune animals brought into the Southern states rendered efforts to develop cattle production through improved breeding and management useless. Before eradication, CFT infestation alone reduced the weight of a 500 kg steer by 100 kg in a year and reduced milk production of dairy cattle by 25% (Agricultural Research Service, 1968).

The common name "cattle tick" was derived in part from the devastating economic impact these parasites had on cattle, and in part from their adaptation to this domestic host (Hoogstraal and Aeschlimann, 1982). However, new parasitic relationships have evolved as CFT invaded various parts of the world. Non-bovine hosts of these parasites include buffalo, ox, horse, mule, donkey, sheep, goat, dog, swine, antelope, and several species of deer (Cooley, 1946; Barré et al., 2001; Ghosh et al., 2007; Cançado et al., 2009; Pound et al., 2010). Ticks spend most of their life in the environment off the host in the egg and larval stages (Figure 4), a period that may last 6–9 months depending on microclimates produced by vegetation communities and rainfall (Newell and Daugherty, 1906; Cotton and Voorhees, 1911; Graybill, 1911). Larvae quest on vegetation for, and eventually attach to a passing host, seek a predilection site to attach and then blood feed. The parasitic portion of their life is completed on the same animal. Therefore, CFT have a one-host life cycle (Figure 4). Blood-fed larvae molt into nymphs and the feeding and molting process is repeated to produce male and female adult CFT. Adults mate and gravid, blood-engorged females detach, drop, and oviposit approximately 3,500 eggs in suitable microclimates at the soil-vegetation interface before dying (Davey et al., 1994). The parasitic phase can be completed over 3–4 weeks providing a substantial period for tick dispersal by foraging animals or human transport of infested animals (Walker et al., 2003). Between three and six generations are produced per year depending on tick species and climatic variables of rainfall and temperature (Mount et al., 1991). In wet-dry season climates, populations generally tend to increase during rainy periods and decline during dry periods (McCulloch and Lewis, 1968; Rawlins, 1979; Daynes and Gutierrez, 1980; Cardozo et al., 1984). Additional information on CFT can be accessed through the TickApp[1] developed to provide educational information on tick biology, ecology, associated diseases, prevention, protection, control, and management.

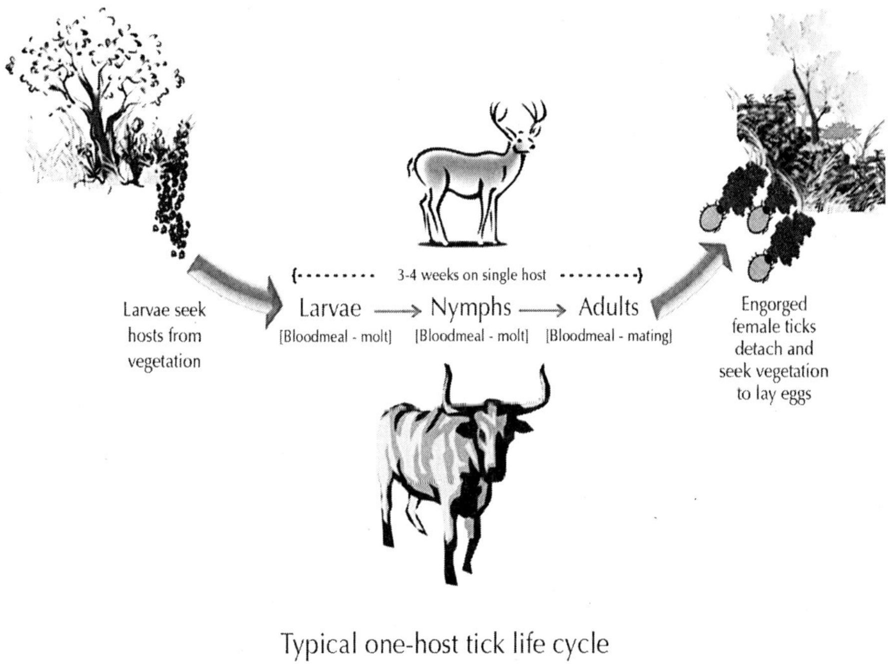

Figure 4: One-host life cycle evolved by cattle fever ticks that comprises an on-host phase where larvae and nymphs sequentially blood feed and molt, and adults feed and mate, and an off-host phase in which mated engorged females drop from the host, oviposit, and die. Larvae hatching from the eggs initiate the next generation. In the absence of hosts, the off-host phase may constitute 87–93% of the life cycle's duration. Used by permission (Teel et al., 2011; http://tickapp.tamu.edu).

HISTORICAL OVERVIEW OF ERADICATION EFFORTS

Early workers observed that separation of CFT from cattle effectively prevented transmission of babesiosis pathogens, and eventually reduced or eliminated tick populations (Curtis and Francis, 1892; Newell and Daugherty, 1906). Tick elimination tactics included cultural practices and evolved with the development of the plunge dip vat to treat cattle with acaricides (Francis, 1894; Graham and Hourrigan, 1977). Regulations for CFT eradication used surveillance to define the spatial boundaries of infestations and quarantine of cattle and premises to implement treatments. Two strategies were employed for tick elimination. A "pasture vacation" strategy first removed ticks from infested cattle by acaricide dipping allowing removal of animals to alternate non-infested premises, then leaving the infested premises vacated for 6–9 months for tick larvae to perish without a host. A "systematic dipping" strategy treated 100% of cattle by dipping at 7–14 day intervals for the 6–9 month period. Acaricide concentrations set to kill 99% of the ticks and effective dip vat management stopped cattle tick development. These procedures required knowing the exact cattle inventory at every dipping session and assuring all animals were confined inside fenced premises. Variations on systematic dipping included double fencing, cross fencing, overstocking, and rotations to improve pasture coverage and increase tick-host contact, in effect "sponging" ticks from the premises (Gray et al., 1979). The CFTEP was declared successful in 1943, and the last outbreak of bovine babesiosis in the U.S. occurred in 1949 (Malone, 1989).

Surveillance and detection were critical to identify and contain CFT incursions as well as to assess success of operational tactics. Probabilities of detecting infested animals or premises is the product of many interacting abiotic and biotic factors influencing spatial and temporal distributions of CFT (Palmer et al., 1976; Teel et al., 1997, 2003). When the resident population is low or widely dispersed due to weather conditions, early stage of establishment, or other factors, the probability of detection is low. Continuous active surveillance by well trained personnel and improvements in detection sampling methods and technologies are essential.

IMPACTS OF GLOBAL CHANGE ON OPERATIONS OF THE CATTLE FEVER TICK ERADICATION PROGRAM

Two spatial and climatic characterizations are relevant to the discussion and interpretation of CFTEP challenges centered in the Tamaulipan scrubland biome, defined as a region whose climate produces similar climax associations of flora and fauna. The Köppen–Geiger climate classification system has recently been updated (Rubel and Kottek, 2010) and provides a common global system for interpretation. A nomenclature for ecological regions, denoted "ecoregions," was developed in the U.S. to identify similarities in type, quality, and quantity of environmental resources as a spatial framework for ecosystem research, assessment, and management (Bryce et al., 1999). This system incorporates variation in biotic and abiotic patterns and composition including geology, physiography, vegetation, climate, soils, land-use, wildlife, and hydrology. Ecoregion names for Texas as defined byCorrell and Johnston (1970) will be used in this discussion. We recognize that climate classification continues to be refined through more robust algorithms and this has implications on our understanding of the ways global change shapes the landscape epidemiology of CFT (Rubel and Kottek, 2010; Cannon, 2012). Major aspects related to global change impacting CFTEP operations in the permanent quarantine zone involve shifts in regional ecology, including wildlife fauna as alternative hosts for CFT. Potential future risks include: climate change, human population impacts on land-use and fractionation, acaricide resistance, and infection of wildlife with strains of *B. bovis*and *B. bigemina* pathogenic to cattle.

Shifts in Regional Ecological Characteristics

The Rio Grande marks the U.S.–Mexico boundary in south Texas and this river flows according to rainfall and reservoir containment as it makes its way toward the Gulf of Mexico. With adequate flow, it provides a physical barrier to slow livestock and wildlife movement between countries, while during drought low-water levels permit increased animal and human movement. The river and CFT quarantine zone cross the Tamaulipan scrub biome (Figure 1B). Based upon the updated Köppen–Geiger climate classification (Rubel and Kottek, 2010), the biome climate includes four categories (Figure 1B). The majority of the biome is classified as an arid-steppe (BSh), which changes to a temperate classification (Cfa) in Texas that extends throughout the historical range of CFT in the southern U.S. (Figure 1A), whereas in Mexico it includes small areas with climate generally classified as arid desert (BWh), or temperate (Cwa). The region is comprised of four ecoregions including Gulf Prairies and Marshes, South Texas Plains (also known as Rio Grande Plains), Edwards Plateau (north east of the river and quarantine line), and the Tamaulipan brushlands (south west of the river) (Figure 1C; Correll and Johnston, 1970). Elevation ranges from sea-level to 1000 feet near the most inland locations. Annual rainfall may be 30–50 inches along the Gulf Coast, but declines to 14–16 inches in the inland reaches of the South Texas Plains-Edwards Plateau. Along the Gulf Coast summers are long, hot and humid (subtropical; Archer et al., 1988) and winters are very mild (>300 frost free days), while progressively more inland locations have hotter, drier summers, and winters subject to more freezing temperatures (Correll and Johnston, 1970). The region is subject to the vagaries of tropical storms from the Caribbean and Gulf of Mexico, as well as periodic and sometimes extended periods of drought. An overlay showing the distribution of CFT species intercepted along the quarantine line reflects the distribution of *R. microplus* in the higher rainfall and more humid areas of the Gulf Prairies and Marshes and South Texas Plains, and the distribution of *R. annulatus*in the drier inland reaches of the South Texas Plains-Edwards Plateau (Figure 1C; adapted fromLohmeyer et al., 2011). Both species have been intercepted in the narrow area near the middle of the quarantine line. Retrospective climate evaluations found suitable habitat for *R. microplus* along Gulf and Atlantic coastal states that included a narrow area at the southern end of Texas (Estrada-Peña et al., 2005). The derived variables of vapor pressure deficit, evaporation, and total rainfall were the principal determinants explaining variation among the 5-year periods of high suitability. Though no climate or habitat suitability studies have been conducted for *R. annulatus*, it is expected that this species could re-establish throughout its original range. A recent risk assessment of hypothetical CFT outbreaks in the temperate zone (Cfa; Figure 1B), east of the South Texas Plains, concluded that significant additional infrastructure and personnel would be needed to meet operational demands, and treatment costs and production losses to producers would likely exceed the economic viability of their operations under current tick elimination options (Anderson et al., 2010).

Descriptions of early landscapes on what is today the Tamaulipan scrubland were of vast grasslands with woody plants limited to sites such as water ways and drainages (Scifres, 1980, 1985). Natural fires, cyclical wet and dry periods, and infrequent grazing by migratory herbivores (e.g., bison, antelope, and deer) minimized overgrazing, and all contributed to the natural management shaping this grassland ecology. Eventual settlement of the area brought about cessation of natural fire and fencing with more intense and long-term grazing by confined herbivores (e.g., cattle, horses, sheep, goats), the combination of which permitted woody plants to expand from their refuges and invade the expanse (Archer et al., 1988). Many areas of South Texas and Northern Mexico eventually gained a continuous cover of brush mottes comprised of large central primary species, such as honey mesquite, *Prosopis glandulosa*, and as many as 15 secondary species in the community. The anthropogenic alteration of the grassland resulted in brush invasion and the transformation to scrubland. Brush encroachments decreased forage production, the primary resource for rangeland cattle (Scifres, 1980), while concurrently providing more favorable habitat for tick survival (Teel, 1991). Consequently, tactics to control brush were developed into integrated brush management strategies to optimize forage production on fertile sites (Scifres and Hamilton, 1985), and these strategies would ultimately include management for wildlife species as well as cattle (Inglis, 1985). Another pathway for the impact of global change is the facilitative ecological interaction between stands of the non-native giant reed (*Arundo donax*) and CFT. The giant reed is an invasive weed that consumes economically significant amounts of water in Mexico and the U.S. Giant reed stands along the Rio Grande were shown to provide abiotic and biotic conditions that are favorable for CFT survival (Racelis et al., 2012). Cooperation between the U.S. and Mexico resulted in a control program to control the giant reed using biocontrol agents (Goolsby et al., 2011), which is likely to alleviate complications this invasive weed presents to the CFTEP.

Efforts to improve rangeland forage production also led to the introduction of drought tolerant buffelgrass, *Cenchrus ciliaris*, from Africa. Rangeland vegetation communities including brush mottes that provide canopy and grasses and forbes of interstitial zones influence the microclimate at the soil-vegetation interface. Spatial distribution of habitats with optimal microclimates interacts with host-landscape behavior to disperse and sustain CFT populations. Management practices for brush control, forage production, grazing, drought, and wildlife have all been developed for land owners/managers engaged in cattle or wildlife production systems in the South Texas Plains. The USDA, Natural Resource Conservation Service, has offered financial and technical assistance for land owners/managers in 17 Texas counties impacted by CFT since 2009. The program funds practices that control brush and improve fences, animal handling facilities, water distribution, and wildlife management[2].

Land-Use and Fractionation

Landscape fragmentation resulting from changes in vegetation cover, property size, use of natural resources, and wildlife diversity and abundance is acknowledged to impact the ecology of tick-borne disease systems, including Lyme disease (Brownstein et al., 2005) and Crimean-Congo hemorrhagic fever (Estrada-Peña et al., 2010). Two CFT outbreaks in 2009 provide a case study in contrasting land-use and fractionation and the impact of a growing human population. An outbreak involving *R. annulatus* occurred in the northern portion of the quarantine zone in country largely devoted to cattle production on large pastures (typically 500–2000 acres) where white-tailed deer (WTD) and human populations are relatively low. In contrast, an *R. microplus* outbreak in the southern portion was characterized by many landowners having small parcels of land (typically 50–200 acres), a comparatively high human population with widely variable land-use and a high concentration of WTD estimated to be one animal per four acres. The South Texas Plains ecoregion (approx. 20 million acres) is comprised of 67% native pasture (rangeland), 10% improved pasture, 4% wildlife management, 13% cropland (dryland/irrigated), and the remainder in a variety of uses (Gilliland et al., 2010). Crop and citrus enterprises are aggregated in the Rio Grande Valley at the very southern portion of the permanent quarantine zone, while the vast majority of land-use to the north is cattle and wildlife on rangeland and pasture. Rural land prices and sales in the South Texas Plains steadily increased from 1990 with sales peaking in 2004 and prices peaking in 2008, before economic recession brought about declines in both parameters (Gilliland et al., 2010). Over recent decades landowners-managers have integrated predominantly cow-calf operations with native and/ or exotic game operations to diversify economic enterprises (Inglis, 1985). Increasing fragmentation of farms and ranches is resulting in rural property size decline driven by social and economic factors including profitability of agricultural enterprises, environmental concerns, and intergenerational land transfers. An evaluation of rural property patterns in Texas (Kjelland et al., 2007) found the strongest predictor of land fragmentation to be non-agricultural value. Changes in land-use and fragmentation are expected to continue to alter the ecology of CFT and the risks associated with bovine babesiosis. The operational impact of these changes on the CFTEP program is extended to the regulatory logistics and public relations associated with the elimination of outbreaks in smaller premises where livestock and wildlife ranching is now practiced. This requires a renewed effort to engage and raise awareness of program goals and procedures among diversified livestock producers.

WILDLIFE AS ALTERNATIVE CATTLE FEVER TICK HOSTS

White-tailed Deer

The deer *Odocoileus virginianus*, commonly known as WTD, was abundant and widespread prior to pioneer settlement (Young, 1956; Rue, 1978). Substantial WTD populations were a resource for meat and hides, and these became significant objects of trade and commerce. One central Texas trading house is reported to have baled and shipped 75,000 deer hides between 1844 and 1853 (Young, 1956). By the late 1800's WTD were at their lowest in many regions (Rue, 1978). The bovine-centered CFT eradication campaign made steady progress until the 1930's when deer infested with*R. microplus* were deemed responsible for persistent infestations in Florida (George, 1990). Though controversial, WTD population reductions were conducted in six counties after other tactics failed, and ticks were subsequently eliminated (George, 1990; Strom, 2009). It is noteworthy that *R.annulatus* was first described from specimens collected from WTD, then known as *Cervus virginianus*, in Florida (Say, 1821).

The esthetic and recreational hunting value of WTD remained, and conservationists working with state-federal agencies re-established and conserved WTD populations throughout the U.S. In the Llano Basin of the Edwards Plateau ecoregion (Figure 1C), an area of approximately 525,000 acres, the average deer density in 1954 was 14.4 per 100 acres, ~75,600 animals, and by 1961 deer density increased to 18.9 per 100 acres with a census of 99,750 animals (Rue, 1978). Successful WTD population growth in Texas through the 1960's was influenced by three factors: beneficial cover and browse was provided by brush vegetation communities; a more defined hunting season was established with conservation goals; and, elimination of the primary screwworm, *Cochliomyia hominovorax* (Rue, 1978). The Texas population in 2004 was estimated at 3.8 million animals (Pound et al., 2010). Range expansion for the Texas WTD subspecies included the Tamaulipan brushlands of Mexico where they are managed for overabundance today (Taylor, 1956; Martinez and Hewitt, 2001).

Most WTD responded to bi-weekly cattle gathering and dipping activities on a CFT infested ranch by temporarily moving to adjacent premises, then returning to the infested and quarantined pastures after activities subsided, while a small number moved into new home ranges (Hood and Inglis, 1974). The behavior of WTD on landscapes has primarily been investigated for deer management and conservation goals (Felix et al., 2007). Robust modeling and field studies are needed on the interactions of cattle and deer (and/or other wildlife species) behavior on tick-infested landscapes to define dynamic relationships of host diversity on tick dispersal and maintenance and to improve tactics for tick suppression and elimination as this aspect has been identified as a gap in our ability to develop effective control strategies to control parasites affecting livestock and wildlife (Morgan et al., 2004). The physiology and grooming behavior of WTD reduced both the number of *R. annulatus*successfully feeding and their subsequent fecundity (Cooksey et al., 1989). Nevertheless, the frequency of both CFT species infesting WTD in both Texas and Mexico increased (Gray et al., 1979;George, 1990). Movement of tick-infested WTD from Mexico across the Rio Grande is often the suspect, or confirmed source of infestations in Texas. Expanses of game fencing were erected as physical barriers to deer, and the previously successful bovine-centric eradication strategies were challenged. Infested premises where the pasture vacation option was implemented often resulted in prolonged infestations and quarantines suggesting WTD were capable of sustaining tick populations in the absence of cattle (George, 1990). Of the two CFT species, *R. microplus* is known to sustain populations on deer in the absence of cattle (Kistner and Hayes, 1970; George, 1990). Reliance on cattle dipping as a means of sponging ticks from infested land (Gray et al., 1979), and the implementation of WTD-specific treatment tactics have been integrated into the CFTEP (George et al., 2008; Pound et al., 2010). Efforts to eradicate CFT infesting alternative wildlife hosts such as WTD present a challenging situation for the CFTEP. The process of issuing quarantines and treating livestock and deer is a significant financial burden on federal and state agencies that also impacts livestock producers. Estimates indicated that the expense to systematically treat every cattle could average $250 per head and lead to an 80% decline in net cash farm income (Anderson et al., 2010). During 2011, approximately 9% of total CFTEP expenditures were allocated to CFT mitigation for deer. Future challenges include optimal integration of these tactics and other emerging technologies as discussed below in adaptable ways to different and changing landscape types for environmentally sound and sustained CFT eradication.

Exotic Ungulates

Interests in foreign wildlife conservation, recreational hunting, and marketing exotic meats provided incentives for private landowners to import exotic ungulates and develop management programs. Initial importations date to the 1920's, and today Texas has more species and greater numbers of exotic game than any other state in North America (Mungall and Sheffield, 1994). Statewide surveys conducted by the Texas Parks and Wildlife Department in 1963 estimated 13 species established and 13,000 animals, and by the final survey in 1996 estimated 76 species and 190,000 animals. These surveys indicate approximately 2/3 are confined by high game fencing and 1/3 are free-ranging. The most successful introductions have been of hoof stock, most originating from Asia. The success of these animals in Texas is due in part to the similarity of respective foreign climates and lack of predators.

The South Texas Plains and Edwards Plateau ecoregions (Figure 1C) have the highest concentration of exotic animals adjacent to the international boundary and permanent quarantine zone (Mungall and Sheffield, 1994). Axis (*Cervus axis*), fallow (*Cervus dama*), and sika deer (*Cervus nippon*), nilgai (*Boselaphus tragocamelus*) and black buck antelope (*Antilope cervicapra*), mouflon (*Ovismusimon*), and aoudad sheep (*Ammotragus lervia*) comprise the most abundant species, and the range of several of these species are known to extend into Mexico. Exotic species discovered with infestations of either *R. microplus* or *R. annulatus* include nilgai, aoudad sheep, wapiti (*Cervus canadensis*), red deer (*Cervus elaphus*), fallow deer, and axis deer (Sheffield et al., 1983; Pound et al., 2010). Like *R. microplus,* the original home of nilgai is the Indian subcontinent. Wapiti and red deer were introduced in Mexico where free-ranging populations live along the border with south Texas (Gallina and Escobedo-Morales, 2009); there are anecdotal reports from Mexico of heavy *R. microplus* infestations in red deer. While the suitability of other exotic hoof stock as potential hosts for CTF is unknown, the adaptability of *R. microplus* to new wildlife hosts has been demonstrated in Brazil and New Caledonia (Cançado et al., 2009; De Meeûs et al., 2010).

Invasive feral swine (*Sus scrofa*) are found throughout the southern region and they have invaded the 10 ecoregions of Texas (Taylor et al., 1998; Rollins et al., 2007; Campbell and Long, 2009). The Texas feral swine population has recently been estimated to be at least 2 million animals (Campbell et al., 2008), and their success credited to high reproductive rate, omnivory, and adaptability (Mungall and Sheffield, 1994). They are significant pests in rural and urban environments, yet also considered valuable for recreational hunting (Rollins et al., 2007). Seven non-CFT species of Ixodid ticks have been found parasitizing feral swine in Texas ecoregions east and north of the border area (Sanders, 2011; Schuster, 2011). Six tick species are three-host ticks and one species, *Dermacentor albipictus,* is a one-host tick. The potential for feral swine to serve as a host for CFT is raised by the recovery of adult *R. microplus* from domestic pigs in Bangladesh (Islam et al., 2006; Ghosh et al., 2007). Additionally, feral swine create frequent breaches in game fences compromising barrier integrity to wildlife containment

(Bodenchuk, 2010). Anthropogenic interests in native and exotic wildlife add host diversity for *R. microplus* and *R. annulatus* in this region, which provides new avenues for tick dispersal and maintenance. Future efforts to eliminate CFT in outbreaks involving exotic hosts will require integration of such factors as habitat use modification, home range management, nutrition, behavior, and the expertise of wildlife specialists to develop best practices (Figure 3).

CLIMATE VARIABILITY AND CHANGE AS DRIVERS OF CFT OUTBREAK SURGES

Solar cycles and global climate influence the dynamics of ecosystems at the macro level (Lean, 2010). A glance at global and national weather/climate events for 2011 provides context for the pathways of global change impacting CFTEP operations. Globally, 2011 was the second wettest year on record since 1900 and this was accompanied by the warmest annual above-average temperatures registered in the southern U.S. The drought in northern Mexico was ranked ninth in a list of top global climate/weather events and Texas experienced its driest year on record in 2011 (National Climatic Data Center, 2012).

Climatic factors limit the geographic distribution of tick species and consequently the epidemiology of tick-borne diseases is also susceptible to the influences of climate (Sonenshine and Mather, 1994;Gray et al., 2009). Where they are endemic, the abundance of tick species is influenced by changes in temperature and moisture. Climate variability is one of the drivers for the emergence and re-emergence of tick-borne diseases (Lafferty, 2009; Randolph, 2010).

In addition to solar cycle activity, the interplay of oceanic and atmospheric processes results in cyclical phenomena that impact regional climates (Gurney and Sapiano, 2006). Examples of such phenomena include the El Niño/Southern Oscillation (ENSO), the Accumulated Cyclone Energy Index (ACE; Enfield and Cid-Serrano, 2010), and other oceanic indicators, which are linked to indices used to forecast weather conditions. Global environmental events influence the population ecology of arthropod-vectors and vector-borne diseases

(Baylis and Githeko, 2006; Linthicum et al., 2008;Anyamba et al., 2012). Anomalies in these phenomena alter the ecological balance, which can result in outbreaks of pests and vector-borne diseases. A warning system based on temporal and spatial predictions of Rift Valley fever outbreak activity derived from analysis of ENSO-related climate anomalies measured through remote sensing of environmental parameters facilitated disease outbreak and mitigation activities (Anyamba et al., 2009). The ENSO, for example, was found to influence CFT populations in North America (Estrada-Peña et al., 2005). Areas in the southern U.S. are subject to sudden changes in habitat suitability for the CFT that are linked to ENSO. This trend highlights a risk for the CFTEP as divergent changes in climate can augment the threat of invasion by*R. microplus* or *R. annulatus* (Estrada-Peña and Venzal, 2006).

At the regional scale, it is noted that the number of tick events recorded within the permanent quarantine zone and in the free zone of south Texas covering fiscal years 1959 through 2011 shows distinct, recurring surges of CFT activity (Figure 5A). By averaging the first (1960–1980) and second (1990–2011) surges, a characteristic tick cycle was revealed (Figure 5B). Significantly, the rapid onset of the surge and equally rapid subsidence, or recovery, suggests that this is a representative and recurring dynamic of CFT outbreaks in this region.

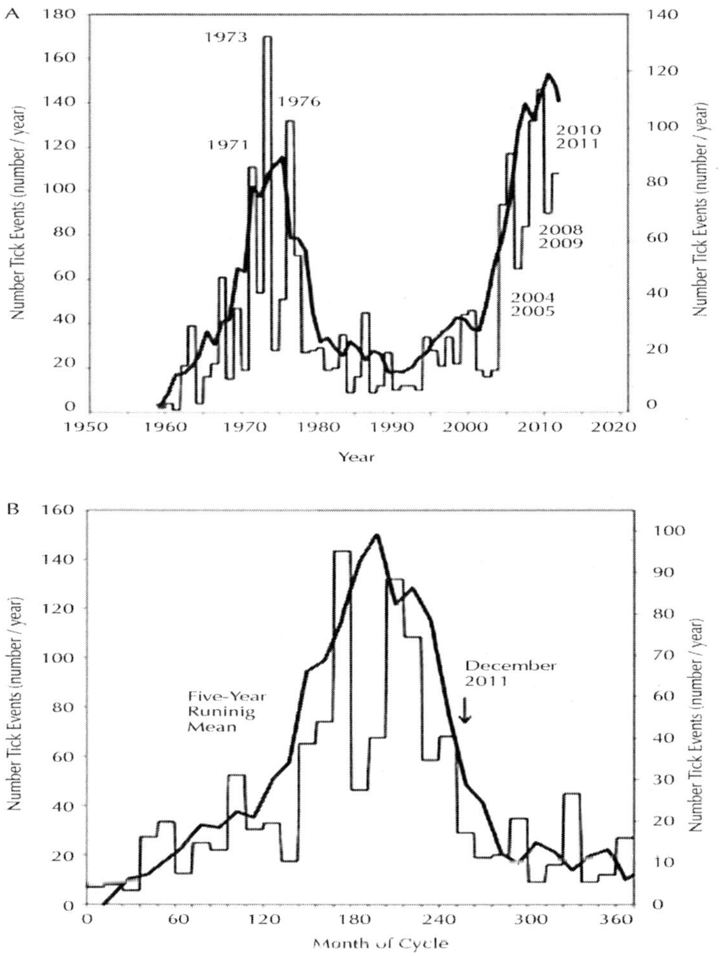

Figure 5: (A,B) Pattern of tick surges in quarantine + free zones of south Texas, 1959–2011, with 5-year running mean (heavy line): (A) yearly level of tick events, showing major years of outbreak in 1970 and 2000 decades; (B)averaged surge periods of 1959–1989, and 1991–2011, showing sharp onset and subsidence of characteristic surge dynamic. December 2011 identifies last sample date and expected drop of tick levels over following year.

Fast Fourier Transform analysis was applied to test the hypothesis of recurring cycles of CFT outbreak activity (SeaSolve, 2004). In addition to identifying principal, recurring CFT events, Fast Fourier Transform analysis enables the recognition and quantification of embedded patterns, which can be important clues to other factors driving CFT population surges such as solar, oceanic, and/or precipitation cycles. Figures 6A–C show a dominant interval lasting 35 years for increased CFT activity in the permanent quarantine + free zone, 33 years in the permanent quarantine zone, and 36 years for outbreak activity in the free zone, respectively. In the latter, the pattern appears simple and likely the result of CFT infestations spreading beyond the permanent quarantine zone that includes a lag interval of about three years. It appears that other factors are operating in the permanent quarantine + free, and permanent quarantine zones (Figures 6A,B). Beyond annual variations, there seems to be a pattern that might include a 4–6 or 8.5 years cycle, which could repeat approximately every 17 years. This observation suggests the operation of ENSO effects with recurrences at those frequencies. It is hypothesized that the 33–36 year interval reflects the influence of the ACE operating at a multidecadal frequency. These spectral intervals are early approximations at this point, and it is recognized that the Fast Fourier Transform analysis is based on only two surges of CFT activity with the most recent surge apparently still in process.

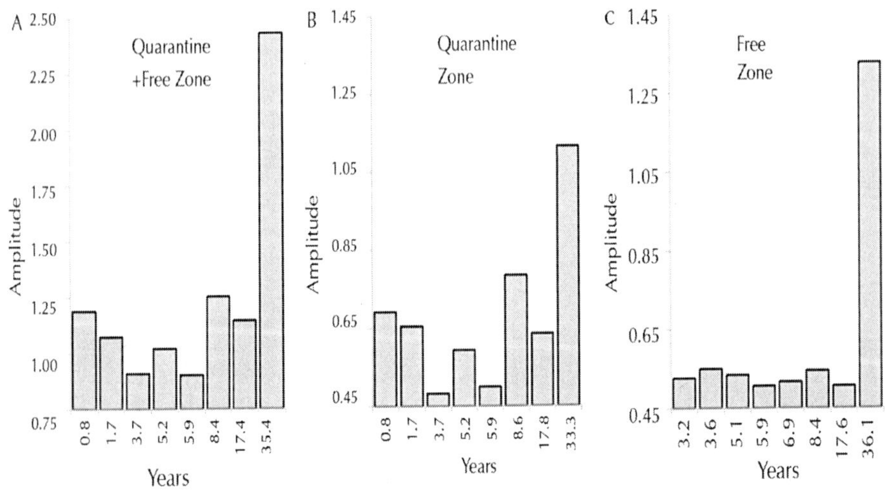

Figure 6: (A–C) Fast Fourier Transform analysis showing strength (amplitude) of main and sub-frequencies (years) of tick outbreak patterns: (A) Quarantine + Free Zone; (B) quarantine zone; (C) Free zone.

A reconstruction of the wavelet analyses as shown in Figure 7 depicts how the results of the Fast Fourier Transform analysis can be applied to forecast future surges of CFT activity in the region. Should the pattern of 1959–2011 repeat itself, one could anticipate a third surge of tick activity in the 2040 decade; uncertainty on its timing and severity needs to be resolved through a better understanding of oceanic signals driving high rainfall intervals in the region, as explored below. Assessing future impacts of climate oscillations and climate perturbations, from tropical hurricanes to acute drought, by modeling normal and extreme conditions would better inform regulatory officials and policy makers on operational needs and regulatory changes as suggested for other disease systems influenced by global change (Garner et al., 2007; Woolhouse, 2011; Anyamba et al., 2012). In particular, strategic forecasting by the CFTEP would enable the adaptation of integrated protocols involving the use of different technologies to minimize the number of, and expedite the elimination of outbreaks. Examples of emerging technologies for integrated eradication are mentioned in the following section.

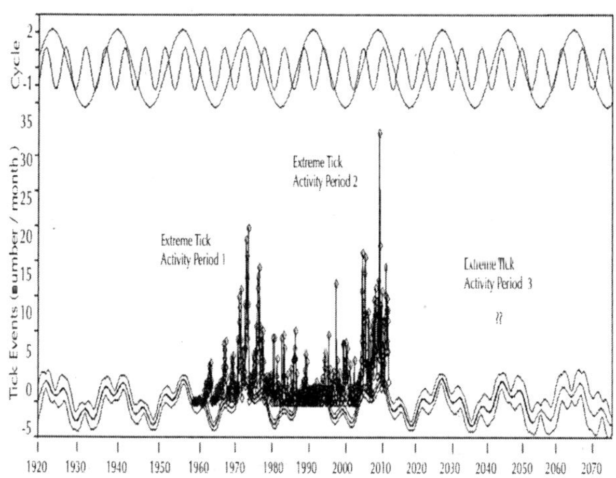

Figure 7: Fast Fourier Transform wavelet analysis of tick events in Quarantine + Free Zone. Two main cycle frequencies are shown at top of graph. Tick outbreak data (1959–2011) on which analysis is based is shown in vertical line and point format. Tick cycles are backcast to 1920 and forecast to 2077 with 90% confidence limits. A period of increased CFT outbreak activity is posited to occur during the 2040 decade.

The field data show the first surge peaked in 1973 when 112 of 170 outbreaks were recorded in the free zone. It took 6 years to reduce the total number of outbreaks at a significant cost for producers and the CFTEP and TAHC, but CFT incursions into the free zone continued to be detected until 1990. The second surge perhaps peaked in 2009 and it appears to be subsiding now. Another *a priori*observation suggests that there is a lag interval where outbreaks in the permanent quarantine zone increase before a transient invasion of the free zone occurs. Thus, the historical record documents recurring, distinct periods of CFT outbreak activity in the south Texas–Mexico border region. The likelihood of recurring episodes is supported by statistical analysis and reconstruction of cycle frequencies in the CFT record. Climate warming was identified as a trend that can alter habitat suitability for *R. microplus* in the Americas including the southeastern U.S. (Estrada-Peña, 2001).

A shift in mean temperature at the global and century scale is primarily a phenomenon of changes at high latitudes of the northern and southern hemispheres. Mean temperature changes in specific regions such as south Texas may or may not coincide with this global trend. The regional trend, for example, decreased from 1895 through about 1975, but then warmed to levels tracking the global increase. Of note, the temperature increases since 1975 occurred in all south Texas NOAA Climate regions from the coast (Region 10) inland to NOAA Climate Region 5, west of the Big Bend border area along the Rio Grande. Moreover, the high plains of the U.S. continental interior, i.e., Platte Drainage, CO, also showed this post-1975 warming pattern, as did Caribbean Region sea surface temperatures. Hence, it appears that a widespread regional temperature trend consistent with, and possibly driven in part by changing global conditions could be influencing CFT outbreak activity, especially in the case of the tropically adapted *R. microplus* as suggested by Estrada-Peña (2001).

Because relative humidity is an important determinant of tick survival (Estrada-Peña, 2001; Corson et al., 2004; Estrada-Peña and Venzal, 2006), we probed for a connection between outbreak surges and long-term regional rainfall. We also queried the possibility that ocean indicators such as the ACE not only drive the rainfall patterns in south Texas (Enfield et al., 2001), but offer an advanced signal of pending major precipitation shifts. The analysis of rainfall in the quarantine + free zones revealed two marked wet periods when the 10-year running precipitation average was applied (Figure 8A). The 10-year averaging was required because any monthly pattern was masked by especially high variability and no apparent regression trend. Notably, these prolonged wet intervals coincide closely with the two CFT outbreak surges (Figure 8A). Moreover, the high rainfall interval appears driven by the ACE, which is a measure of North Atlantic and Caribbean Region hurricane severity that brings rain into south Florida and south Texas. Since 1975, the lag between the intervals of high rain periods in south Texas and increase in ACE is ~3–5 years (Figure 8B). Hence, a simple climatic algorithm to track outbreak surges that offers the additional possibility of anticipating such surges years in advance is presented here. The Atlantic Multidecadal Oscillation, thought to correlate with and drive the ACE (Enfield and Cid-Serrano, 2010), also provides advance warning of changes in the south Texas rainfall.

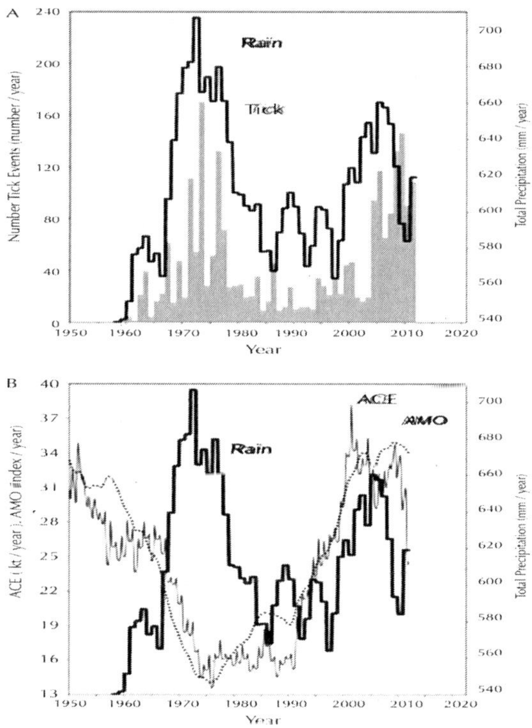

Figure 8: (A,B) Comparison of annual total precipitation (rain, heavy line) in south Texas, 1959–2011 to: (A) number of tick events in quarantine + free zones (light gray); (B) magnitude of the annual Accumulated Cyclone Energy Index (ACE) and Atlantic Multidecadal Oscillation (AMO). The AMO at 1/(45*x* + 25) of scale shown. No lags are applied.

ACARICIDE RESISTANCE RISK AND TECHNOLOGICAL OPPORTUNITIES FOR INTEGRATED ERADICATION

The CFT *R. microplus* is ranked as the sixth most pesticide-resistant arthropod in the world (Whalon et al., 2008). This situation reflects the historical reliance on chemical acaricides, which are pesticides that kill ticks, to control and eradicate CFT. Arsenical dips were introduced in 1911 and they were used by the CFTEP until the 1960s (Graham and Hourrigan, 1977). Organophosphate acaricides replaced arsenical dips in 1967, and in 1971 the CFTEP started using coumaphos to dip cattle and spray horses. Additionally, an injectable macrocyclic lactone acaricide formulation is used on a need-basis by the CFTEP during outbreaks where dipping alone presents difficulties for eradication. The two technologies employed to eliminate CFT infesting WTD are ivermectin-medicated corn and topical permethrin self-treatment using a bait station fitted with the "2-poster" adapter (Pound et al., 2010). Corn medicated with ivermectin was used to eradicate long-term CFT infestations maintained by wapiti and WTD (Pound et al., 2009).

In Mexico, organophosphate acaricides were used for the national tick eradication program between 1974 and 1984 (Trapaga, 1989). Pyrethroids and amitraz were authorized for CFT control in Mexico since 1985 (Rosario-Cruz et al., 2009). The use of amitraz gained popularity after CFT became refractory to treatment with organophosphates and pyrethroids, but resistance to amitraz was eventually reported in 2001 (Domínguez-García et al., 2010). The ticks have also evolved resistance to fipronil and multiple resistance to organophosphates, pyrethroids, amitraz, and ivermectin in Mexico (Miller et al., 2008; Fernández-Salas et al., 2012). Resistance to organophosphates and pyrethroids has been detected in CFT collected in Texas (Miller et al., 2005, 2007). Surveillance of acaricide resistance in CFT from outbreaks and other sources is an ongoing effort by the Cattle Fever Tick Research Laboratory of the USDA-Agricultural Research Service because this is a constant threat to CFTEP operations.

Anti-CFT vaccination in cattle and wildlife, and the passive administration of systemic acaricides for cattle are two technologies that could be used in an integrated fashion by the CFTEP (Pérez de León et al., 2010). Vaccines against *R. microplus* based on the protective antigen Bm86 were developed and marketed under the names TickGARD (Willadsen et al., 1995), and Gavac (Canales et al., 1997). Their efficacy against that CFT species generally ranges from 10 to 89% (Willadsen, 2008). The level of efficacy reported for vaccines containing a recombinant form of Bm86 varies according to, among other things, the expression system used for vaccine production, characteristics of the tick population targeted, and host factors (de la Fuente et al., 1999; Patarroyo et al., 2002; Sitte et al., 2002;Casquero Cunha et al., 2012). It has been postulated that sequence variation in the *bm86* locus is associated with low susceptibility of certain *R. microplus* populations infesting cattle immunized with commercial Bm86-based vaccines (García-García et al., 1999). Interestingly, the efficacy of Bm86-based vaccines against *R. annulatus* is >99%; this level of efficacy is equivalent to that obtained with commercial acaricides in the absence of resistance. Anti-CFT vaccines based on Bm86 play an important role in integrated tick management systems (de la Fuente et al., 2007; Willadsen, 2008). Strategies combining anti-CFT vaccination and acaricides were shown to reduce the number of acaricidal applications necessary for tick control (de la Fuente et al., 2007), which also extends the useful life of acaricides by delaying the development of acaricide resistance. Gavac is not commercially available in the U.S. This vaccine was 99% efficacious against an outbreak strain of *R.annulatus* (Miller et al., 2012).

The post-genomic era offers the opportunity to take a systems biology-based approach to antigen discovery and vaccine development (Oberg et al., 2011). A web-based repository of *R. microplus*genomic data is helping with the identification and selection of sequences coding for molecules that can be tested as protective antigens (Bellgard et al., 2012). Ecologically based modeling can also be applied to evaluate and adjust various factors to maximize vaccine efficacy against CFT invading the U.S. (Miller et al., 2012). All these approaches are being applied to discover novel antigens that protect cattle in the permanent quarantine zone against *R. microplus* infestation, which could also be used to vaccinate WTD (Guerrero et al., 2012). Carreón et al. (2012) proved the feasibility of protecting WTD against CFT infestation through vaccination. The utility of vaccinating WTD for CFT eradication in south Texas could be tested initially in high-fenced premises. Combining vaccination with the existing CFT treatments for WTD will eliminate their role as reservoirs for the ticks.

As compared to other suitable hosts, populations of CFT realize their full reproductive potential when parasitizing cattle. Although it is likely that CFTEP strategies will remain focused on eliminating CFT infesting cattle, the use of methods applied to wildlife is needed because it appears that pasture vacation represents an anachronism from the time when cattle, horses, mules, and asses were the only hosts of concern (George et al., 2008). Therefore, it is proposed that the combined use of emerging technologies like vaccination of cattle and WTD against CFT and the use of an ivermectin-medicated feed block be explored. The integration of these tools with existing treatment protocols could lessen the economic and operational impacts of quarantines on producers.

RISK FOR WILDLIFE WITH STRAINS OF *B. BOVIS* AND *B. BIGEMINA* PATHOGENIC TO CATTLE

Attempts to infect WTD with the agents of bovine babesiosis were unsuccessful (Kuttler et al., 1972). A study investigating the possibility that WTD might serve as a reservoir of bovine babesiosis revealed that the addition of bovine serum was required for *in vitro* culture of *B. bovis* in WTD erythrocytes (Holman et al., 1993). Serologic evidence and results from PCR tests indicated that WTD in Mexico could be infected with *B. bovis* and *B. bigemina* (Cantu-Covarrubias et al., 2009). Similar evidence exists for WTD in Texas (Holman et al., 2011). However, the data thus far suggests that WTD bitten by CFT infected with *B. bovis* or *B. bigemina* function as diluting hosts because these piroplasms are unable to undergo the asexual portion of their life cycle in WTD and become infective for CFT. The deer may be having a zooprophylactic effect by reducing the prevalence of infected CFT in the environment.

The nilgai is a tropically adapted bovid that poses a risk for the re-emergence of bovine babesiosis in the U.S. because, in addition to its host ability for CFT, this species adapts well to agricultural areas and commingles with cattle in south Texas (Sheffield et al., 1971; Davey, 1993; Leslie, 2008). *Babesia*infection in a nilgai from India was reported (Baviskar et al., 2009). There are approximately 30,000 nilgais along the south Texas–Mexico border and a population in Mexico was polymerase chain reaction-positive for *B. bovis* and *B. bigemina* (Cárdenas-Canales et al., 2011).

PERSPECTIVES

The predominant CFT species at the beginning of the eradication campaign was *R. annulatus*. By comparison *R. microplus* is a more invasive species. Records indicate CFT outbreaks in south Texas are frequently caused by *R. microplus*. The invasion and re-invasion of other parts of the world by *R.microplus* highlight the danger this tick poses to U.S. animal agriculture. Climate change could augment the economic impact of CFT by influencing habitat suitability and shifts in potential distribution of both species. Climate oscillations of wet and dry periods will undoubtedly continue to impact CFT and the preparedness of the CFTEP to respond to these changes. Additional hypothesis-driven research is required to effectively understand and manage the risk of CFT and bovine babesiosis re-emergence in the U.S. The complicated nature of current and future problems driven by global change requires that societal, environmental, scientific, and policy considerations be integrated to develop the most effective and sustainable strategies and goals to prevent re-establishment of CFT and/or outbreaks of bovine babesiosis (Figure 3).

Analyses of climate change effects presented here demonstrate that North Atlantic oceanic indicators, i.e., AMO and ACE, and region-scale rainfall shifts have a marked impact on periodic, CFT outbreak activity in south Texas, and may provide opportunity to develop responsive surveillance strategies. Here, it was documented that the collection and analysis of meaningful data allowing assessment and prediction to identify early warnings of pending changes in CFT outbreak activity is possible. It is recognized that future developments in monitoring and interpreting important ocean signals such as the AMO and ACE will continue to improve predictive power. Ongoing and future efforts to carefully systematize CFT records and put them into electronic format for pre-structured models designed to share with and fully inform decision makers at all levels is justified.

Comprehensive studies are advocated to explore thoroughly climate impacts on CFT, as well as other ticks, to design and integrate best management practices to evaluate ecosystem shifts affecting emergence of tick-borne diseases, and to inform policy makers for rational and effective decision making. This knowledge base could be used as the foundation to create and implement an international CFT monitoring and prediction program for the assessment and analysis of changes in tick distribution, ecology, host utilization, and disease epidemiology to mitigate the economic impact of these ticks and the pathogens they transmit. Such systems have been applied successfully to other arthropod-borne diseases and they could be consolidated in existing databases (Anyamba et al., 2009; Gale et al., 2010; Vesco et al., 2011). Worldwide coordination to adapt and refine prediction systems for ticks and tick-borne diseases will enable the enhancement of public health and veterinary services programs to anticipate, prepare, and respond to these emerging and re-emerging human and animal health threats (Institute of Medicine, and National Research Council, 2009).

ACKNOWLEDGMENTS

Thanks to Sherri Brown and Helen Ramsey for outstanding editorial support and to Otto Strey for graphics support to create Figures 1A–C. Drs. Dan Strickman, John Goolsby, and Samuel Liu provided constructive comments on an earlier version of the manuscript. The research of Drs. Pérez de León, Felix Guerrero, and Robert Miller is supported by appropriated funds for projects 6205-32000-031-00D and 6205-32000-034-00D. Dr. Schuster's research is funded in part through NIFA project TEXR-2009-05759. USDA is an equal opportunity provider and employer. Ms. Johanna Reichen with the Frontiers Production Office is gratefully acknowledged for expert production coordination.

REFERENCES

1. Agricultural Research Service. (1968). *The Fight Against Cattle Fever Ticks*. Hyattsville, MD: U.S. Government Printing Office, 320–594.

2. Aguilar, J. (2011). *Cartel Violence Complicates Tick Eradication Plan. New York Times*. Available at:http://www.nytimes.com/2011/04/17/us/17ttticks.html [accessed April 16, 2011].

3. Anderson, D. P., Hagerman, A. D., Teel, P. D., Wagner, G. G., Outlaw, J. L., and Herbst, B. K. (2010). *Economic Impact of Expanded Fever Tick Range*. Research Report 10-2. College Station, TX: *Agricultural and Food Policy Center, Texas AgriLife Research*.

4. Anyamba, A., Chretien, J. P., Small, J., Tucker, C. J., Formenty, P. B., Richardson, J. H., Britch, S. C., Schnabel, D. C., Erickson, R. L., and Linthicum, K. J. (2009). Prediction of a Rift Valley Fever outbreak. *Proc. Natl. Acad. Sci. U.S.A.*106, 955–959.

5. Anyamba, A., Linthicum, K. J., Small, J. L., Collins, K. M., Tucker, C. J., Pak, E. W., Britch, S. C., Eastman, J. R., Pinzon, J. E., and Russell, K. L. (2012). Climate teleconnections and recent patterns of human and animal disease outbreaks. *PLoS Negl. Trop. Dis.* 6, e1465. doi:10.1371/journal.pntd.00011465

6. Archer, S., Scifres, C. J., Bassham, C. R., and Maggio, R. (1988). Autogenic succession in a subtropical savanna: conversion of grassland to thorn woodland. *Ecol. Monogr.* 58, 111–127.

7. Aubry, P., and Geale, D. W. (2011). A review of bovine anaplasmosis. *Transbound. Emerg. Dis.* 58, 1–30.

8. Barré, N., Bianchi, M., and Chardonnet, L. (2001). Role of Rusa deer Cervus timorensis russa in the cycle of the cattle tick *Boophilus microplus* in New Caledonia. *Exp. Appl. Acarol.* 25, 79–96.

9. Baviskar, B., Ingle, K., Gawande, P., Raut, S., Sirothia, K., and Bhandarkar, A. (2009). First report on occurrence of*Babesia* infection in nilgai *Boselaphus tragocamelus* from central India. *J. Threat. Taxa* 1, 196.

10. Baylis, M., and Githeko, A. K. (2006). *Foresight. Infectious Diseases: Preparing for the Future. T7.3: The Effects of Climate Change on Infectious Diseases of Animals*. London: Office of Science and Innovation.

11. Bellgard, M., Moolhuijzen, P. M., Guerrero, F. D., Schibeci, D., Rodriguez-Valle, M., Peterson, D. G., Dowd, S. E., Barrero,

R., Hunter, A., Miller, R. J., and Lew-Tabor, A. E. (2012). CattleTickBase: internet-based analysis tools and bioinformatics repository of available genomics resources for *Rhipicephalus (Boophilus) microplus*. *Int. J. Parasitol.* 42, 161–169.

12. Bishopp, F. C. (1913). The occurrence of the Australian Cattle Tick and the Brown Dog-tick in Key West, Florida (Acarina, Ixodoidea). *Ent. News* 24, 366–368.

13. Black, P., and Nunn, M. (2009). "Impact of climate change and environmental changes on emerging and re-emerging animal disease and animal production." *Paper Presented at the 77th General Session World Organization for Animal Health, International Committee*, Paris.

14. Bodenchuk, M. J. (2010). *USDA-APHIS-Wildlife Services Feral Hog Project Accomplishments, 2008–2010*. San Antonio, TX: Texas Department of Agriculture.

15. Bram, R. A., George, J. E., Reichar, R. E., and Tabachnick, W. J. (2002). Threat of foreign arthropod-borne pathogens to livestock in the United States. *J. Med. Entomol.* 39, 405–416.

16. Brouqui, P. (2011). Arthropod-borne diseases associated with political and social disorder. *Annu. Rev. Entomol.*56, 357–374.

17. Brownstein, J. S., Skelly, D. K., Holford, T. R., and Fish, D. (2005). Forest fragmentation predicts local scale heterogeneity of Lyme disease risk. *Oecologia* 146, 469–475.

18. Bryce, S. A., Omernik, J. M., and Larsen, D. P. (1999). Ecoregions – a geographic framework to guide risk characterization and ecosystem management. *Environ. Pract.* 1, 141–155.

19. Bulloch, W. (1935). Theobald Smith. *J. Pathol. Bacteriol.* 40, 621–635.

20. Camill, P. (2010). Global change. *Nat. Educ. Knowl.* 2, 49.

21. Campbell, T. A., DeYoung, R. W., Wehland, E. M., Grassman, L. I., Long, D. B., and Delgado-Acevedo, J. (2008). Feral swine exposure to selected viral and bacterial pathogens in southern Texas. *J. Swine Health Prod.* 16, 312–315.

22. Campbell, T. A., and Long, D. B. (2009). Feral swine damage and damage management in forested ecosystems. *For. Ecol. Manage.* 257, 2319–2336.

23. Canales, M., Enriquez, A., Ramos, E., Cabrera, D., Dandie, H., Soto, A., Falcon, V., Rodriguez, M., and de la Fuente, J. (1997). Large-scale production in *Pichia pastoris* of the recombinant vaccine Gavac against cattle tick. *Vaccine*15, 414–422.

24. Cançado, P. H. D., Zucco, C. A., Piranda, E. M., Faccini, J. L. H., and Mourão, G. M. (2009). *Rhipicephalus (Boophilus) microplus* (Acari: Ixodidae) as a parasite of pampas deer (*Ozoctoceros bezoarticus*) and cattle in Brazil's Central Pantanal. *Rev. Bras. Parasitol. Vet.* 18, 42–46.

25. Cannon, A. J. (2012). Köppen versus the computer: comparing Köppen-Geiger and multivariate regression tree climate classifications in terms of climate homogeneity. *Hydrol. Earth Syst. Sci.* 16, 217–229.

26. Cantu-Covarrubias, A., Ortega-Santos, J. A., García-Vázquez, Z., Mosqueda, J., Henke, S. E., and George, J. E. (2009). Epizootiology of *Babesia bovis* and *Babesia bigemina* in free-ranging white-tailed deer in northeastern Mexico. *J. Parasitol.* 95, 536–542.

27. Cárdenas-Canales, E. M., Ortega-Santos, J. A., Campbell, T. A., Garciá-Vázquez, Z., Cantú-Covarrubias, A., Figueroa-Millán, J. V., DeYoung, R. W., Hewitt, D. G., and Bryant, F. C. (2011). Nilgai antelope in northern Mexico as a possible carrier for cattle fever ticks and *Babesia bovis* and *Babesia bigemina*. *J. Wildl. Dis.* 47, 777–779.

28. Cardozo, H., Nari, A., Franchi, M., Lopez, A., and Donatti, N. (1984). Estudios sobre la ecologia del *Boophilus microplus* en tres areas enzooticas del Uruguay. *Veterinaria* 20, 4–10.

29. Carreón, D., Pérez de la Lastra, J. M., Almazán, C., Canales, M., Ruiz-Fons, F., Boadella, M., Moreno-Cid, J. A., Villar, M., Gortázar, C., Reglero, M., Villarreal, R., and de la Fuente, J. (2012). Vaccination with BM86, subolesin and akirin protective antigens for the control of tick infestations in white tailed deer and red deer. *Vaccine* 30, 273–279.

30. Casquero Cunha, R., Pérez de León, A. A., Leivas Leite, F. P., da Silva Pinto, L., Gonçalves dos Santos, A. Jr., and Andreotti, R. (2012). Bovine immunoprotection against *Rhipicephalus microplus* with recombinant Bm86-Campo Grande antigen. *Rev. Bras. Parasitol. Vet.* (in press).

31. Cooksey, L. M., Davey, R. B., Ahrens, E. H., and George, J. E. (1989). Suitability of white-tailed deer as hosts for cattle fever ticks (Acari: Ixodidae). *J. Med. Entomol.* 26, 155–158.
32. Cooley, R. A. (1946). *The Genera Boophilus Rhipicephalus, and Haemaphysalis (Ixodidae) of the New World.* Washington: National Institutes of Health Bulletin No. 187, 1–54.
33. Correll, D. S., and Johnston, M. C. (1970). *Manual of the Vascular Plants of Texas.* Renner: Texas Research Foundation.
34. Corson, M. S., Teel, P. D., and Grant, W. E. (2004). Microclimate influence in a physiological model of cattle-fever tick (*Boophilus* spp.) population dynamics. *Ecol. Modell.* 180, 487–514.
35. Cotton, E. C., and Voorhees, J. F. (1911). *The Cattle Tick as Affected by Climate.* Bulletin No. 94, Knoxville, TN: Tennessee Agricultural Experiment Station.
36. Cumming, G. S., and Van Vuuren, D. P. (2006). Will climate change affect ectoparasite species ranges? *Global Ecol. Biogeogr.* 15, 486–497.
37. Curtis, C., and Francis, M. (1892). *The Cattle Tick: Biology, Preventive Measures.* Bulletin No. 24. College Station, TX: Texas Agricultural Experiment Station.
38. Davey, R. B. (1993). Stagewise mortality, ovipositional biology, and egg viability of *Boophilus annulatus* (Acari: Ixodidae) on *Boselaphus tragocamelus* (Artiodactyla: Bovidae). *J. Med. Entomol.* 30, 997–1002.
39. Davey, R. B., Pound, J. M., and Cooksey, L. M. (1994). Comparative reproduction and nonparasitic development of*Boophilus microplus* and hybridized *Boophilus* ticks (Acari: Ixodidae) under natural field conditions in subtropical South Texas. *Exp. Appl. Acarol.* 18, 185–200.
40. Daynes, P., and Gutierrez, J. (1980). Variations saisonniees de l'activite parasitaire de la tique du betail *Boophilus microplus* (Acari: Ixodidae) en Nouvelle-Caledonie. *Rev. Elev. Med. Vet. Pays. Trop.* 33, 305–310.
41. de la Fuente, J., Almazan, C., Canales, M., Perez de la Lastra, J. M., Kocan, K. M., and Willadsen, P. (2007). A ten-year review of commercial vaccine performance for control of tick infestations on cattle. *Anim. Health Res. Rev.* 8, 23–28.

42. de la Fuente, J., Rodriguez, M., Montero, C., Redondo, M., García-García, J. C., Méndez, L., Serrano, E., Valdés, M., Enríquez, A., Canales, M., Ramos, E., Boué, O., Machado, H., and Lleonart, R. (1999). Vaccination against ticks (Boophilus spp.): the experience with the Bm86-based vaccine Gavac. *Genet. Anal.* 15, 143–148.

43. De Meeûs, T., Koffi, B. B., Barré, N., de Garine-Wichatitsky, M., and Chevillon, C. (2010). Swift sympatric adaptation of a species of cattle tick to a new deer host in New Caledonia. *Infect. Genet. Evol.* 10, 976–983.

44. Domínguez-García, D. I., Rosario-Cruz, R., Almazán García, C., Saltijeral Oaxaca, J. A., and de la Fuente, J. (2010).*Boophilus microplus*: biological and molecular aspects of acaricide resistance and their impacts on animal health.*Trop. Subtrop. Agroecosyst.* 12, 181–192.

45. Enfield, D. B., and Cid-Serrano, L. (2010). Secular and multidecadal warmings in the North Atlantic and their relationships with major hurricane activity. *Int. J. Climatol.* 30, 174–184.

46. Enfield, D. B., Mestas-Nunez, A. M., and Trimble, P. J. (2001). The Atlantic multidecadal oscillation and its relationship to rainfall and river flows in the continental U.S. *Geophys. Res. Lett.* 28, 2077–2080.

47. Estrada-Peña, A. (2001). Climate warming and changes in habitat suitability for *Boophilus microplus* (Acari: Ixodidae) in Central America. *J. Parasitol.* 87, 978–987.

48. Estrada-Peña, A., Acedo, S., Quílez, J., and Del Cacho, E. (2005). A retrospective study of climatic suitability for the tick *Rhipicephalus* (*Boophilus*) *microplus* in the Americas. *Glob. Ecol. Biogeogr.* 14, 565–573.

49. Estrada-Peña, A., Bouattour, A., Camicas, J. L., Guglielmone, A., Horak, I., Jongejan, F., Latif, A., Pegram, R., and Walker, A. R. (2006). The known distribution and ecological preferences of the tick subgenus *Boophilus* (Acari: Ixodidae) in Africa and Latin America. *Exp. Appl. Acarol.* 38, 219–235.

50. Estrada-Peña, A., Vatansever, Z., Gargili, A., and Ergönul, O. (2010). The trend towards habitat fragmentation is the key factor driving the spread of Crimean-Congo haemorrhagic fever. *Epidemiol. Infect.* 138, 1194–1203.

51. Estrada-Peña, A., and Venzal, J. M. (2006). High-resolution predictive mapping for *Boophilus annulatus* and *B. microplus (Acari: Ixodidae)* in Mexico and Southern Texas. *Vet. Parasitol.* 142, 350–358.

52. Estrada-Peña, A. (2008). Climate, niche, ticks, and models: what they are and how we should interpret them.*Parasitol. Res.* 103(Suppl. 1), S87–S95.

53. Felix, A. B., Walsh, D. P., Hughey, B. D., Campa, H., and Winterstein, S. R. (2007). Applying landscape-scale habitat-potential models to understand deer spatial structure and movement patterns. *J. Wildl. Manage.* 71, 804–810.

54. Fernández-Salas, A., Rodríguez-Vivas, R. I., and Alonso-Díaz, M. A. (2012). First report of a *Rhipicephalus microplus* tick population multi-resistant to acaricides and ivermectin in the Mexican tropics. *Vet. Parasitol.* 183, 338–342.

55. Food and Agriculture Organization of the United Nations (FAO). (2008). *Climate Change, Energy and Food – Climate-Related Transboundary Pests and Diseases*. Rome: FAO.

56. Food and Agriculture Organization of the United Nations (FAO). (2011). *World Livestock 2011 – Livestock in Food Security*. Rome: FAO.

57. Francis, M. (1894). *A Device for Dipping Cattle to Destroy Ticks*. Bulletin No. 30. College Station, TX: Texas Agricultural Experiment Station.

58. Gale, P., Estrada-Peña, A., Martinez, M., Ulrich, R. G., Wilson, A., Capelli, G., Phipps, P., de la Torre, A., Muñoz, M. J., Dottori, M., Mioulet, V., and Fooks, A. R. (2010). The feasibility of developing a risk assessment for the impact of climate change on the emergence of Crimean-Congo haemorrhagic fever in livestock in Europe: a review. *J. Appl. Microbiol.* 108, 1859–1870.

59. Gallina, S., and Escobedo-Morales, L. A. (2009). Análisis sobre las Unidades de Manejo (UMAs) de ciervo rojo (*Cervus elaphus* Linnaeus, 1758) y wapiti (*Cervus canadensis* (Erxleben, 1777) en México: problemática para la conservación de los ungulados nativos. *Trop. Conserv. Sci.* 2, 251–265.

60. García-García, J. C., Gonzalez, I. L., González, D. M., Valdés, M., Méndez, L., Lamberti, J., D'Agostino, B., Citroni, D., Fragoso, H., Ortiz, M., Rodríguez, M., and de la Fuente, J. (1999).

Sequence variations in the *Boophilus microplus*Bm86 locus and implications for immunoprotection in cattle vaccinated with this antigen. *Exp. Appl. Acarol.* 23, 883–895.

61. Garner, M. G., Dubé, C., Stevenson, M. A., Sanson, R. L., Estrada, C., and Griffin, J. (2007). Evaluating alternative approaches to managing animal disease outbreaks – the role of modelling in policy formulation. *Vet. Ital.* 43, 285–298.
62. George, J. (2008). The effects of global change on the threat of exotic arthropods and arthropod-borne pathogens to livestock in the United States. *Ann. N. Y. Acad. Sci.* 1149, 249–254.
63. George, J. E. (1989). *The Eradication of Ticks – Cattle Fever Tick Eradication Programme in the USA: History, Achievements, Problems and Implications for Other Countries*. Rome: Food and Agriculture Organization of the United Nations, FAO, 1–7.
64. George, J. E. (1990). Wildlife as a constraint to the eradication of *Boophilus* spp. (Acari: Ixodidae). *J. Agric. Entomol.* 7, 119–125.
65. George, J. E., Pound, J. M., Kammlah, D., and Lohmeyer, K. H. (2008). "Presumptive evidence for the role of the white-tailed deer in the epidemiology of *Rhipicephalus* (*Boophilus*) *annulatus* and *Rhipicephalus* (*Boophilus*)*microplus* (Acari:Ixodidae)," in *VI Seminario Internacional de Parasitologia Animal*, Boca del Rio Veracruz: Instituto Nacional de Investigaciones Forestales, Agricolas y Pecuaries.
66. Ghosh, S., Bansal, G. C., Gupta, S. C., Ray, D., Khan, M. Q., Irshad, H., Shahiduzzaman, M., Seitzer, U., and Ahmed, J. S. (2007). Status of tick distribution in Bangladesh, India, and Pakistan. *Parasitol. Res.* 101, 207–216.
67. Gilliland, C. E., Gunadekar, A., Wiehe, K., and Whitmore, S. (2010). *Characteristics of Texas Land Markets, A Regional Analysis*. Technical Report 1937. College Station, TX: Real Estate Center, Texas A & M University.
68. Goolsby, J. A., Kirk, A. A., Moran, P. J., Racelis, A. E., Adamczyk, J. J., Cortés, E., Marcos García, M. Á., Martinez Jimenez, M., Summy, K. R., Ciomperlik, M. A., and Sands, D. P. A. (2011). Establishment of the armored scale,*Rhizaspidiotus donacis*, a biological control agent of *Arundo donax*. *Southwest. Entomol.* 36, 373–374.

69. Graham, O. H., and Hourrigan, J. L. (1977). Eradication programs for the arthropod parasites of livestock. *J. Med. Entomol.* 13, 629–658.

70. Gray, J., Dautel, H., Estrada-Peña, A., Kahl, O., and Lindgren, E. (2009). Effects of climate change on ticks and tick-borne diseases in Europe. *Interdiscip. Perspect. Infect. Dis.* 2009, 12.

71. Gray, J. H., Payne, R. L., Schubert, G. O., and Garnett, W. H. (1979). Implication of white-tailed deer in the*Boophilus annulatus* tick eradication program. *Proc. Annu. Meet. U. S. Anim. Health Assoc.* 83, 506–515.

72. Graybill, H. W. (1911). *Studies on the Biology of the Texas-Fever Tick*. Bulletin No. 130. Washington: USDA, Bureau of Animal Industries.

73. Guerrero, F. D., Bendele, K. G., Davey, R. B., and George, J. E. (2007). Detection of *Babesia bigemina* infection in strains of *Rhipicephalus* (*Boophilus*) *microplus* collected from outbreaks in south Texas. *Vet. Parasitol.* 145, 156–163.

74. Guerrero, F. D., Miller, R. J., and Pérez de León, A. A. (2012) Anti-cattle tick vaccines: many candidate antigens, but will a commercially viable product emerge? *Int. J. Parasitol.* 42, 421–427.

75. Gurney, R., and Sapiano, M. (2006). *Foresight. Infectious Diseases: Preparing for the Future. S10: State-of-Science Review: Earth Observation*. London: Office of Science and Innovation.

76. Hanson, C., Ranganathan, J., Iceland, C., and Finisdore, J. (2008). *The Corporate Ecosystem Services Review: Guidelines for Identifying Business Risks and Opportunities Arising from Ecosystem Change*. Washington: Meridian Institute, World Resources Institute, World Business Council for Sustainable Development.

77. Harrus, S., and Baneth, G. (2005). Drivers for the emergence and re-emergence of vector-borne protozoal and bacterial diseases. *Int. J. Parasitol.* 35, 1309–1318.

78. Hoberg, E. P., Polley, L., Jenkins, E. J., and Kutz, S. J. (2008). Pathogens of domestic and free-ranging ungulates: global climate change in temperate to boreal latitudes across North America. *Rev. Sci. Tech.* 27, 511–528.

79. Holman, P. J., Carroll, J. E., Pugh, R., and Davis, D. S. (2011). Molecular detection of *Babesia bovis* and *Babesia bigemina* in white-tailed deer (*Odocoileus virginianus*) from Tom Green County in central Texas. *Vet. Parasitol.*177, 298–304.

80. Holman, P. J., Waldrup, K. A., Droleskey, R. E., Corrier, D. E., and Wagner, G. G. (1993). *In vitro* growth of *Babesia bovis* in white-tailed deer (*Odocoileus virginianus*) erythrocytes. *J. Parasitol.* 79, 233–237.

81. Hood, R. E., and Inglis, J. M. (1974). Behavioral responses of white-tailed deer to intensive ranching operations. *J. Wildl. Manage.* 38, 488–498.

82. Hoogstraal, H., and Aeschlimann, A. (1982). Tick-host specificity. *Bull. Soc. Entomol. Suisse* 55, 5–32.

83. Inglis, J. M. (1985). "Wildlife management and IBMS," in *Integrated Brush Management Systems for South Texas: Development and Implementation*. Publication No. 2M-6-85. College Station, TX: Texas Agricultural Experiment Station.

84. Institute of Medicine and National Research Council. (2009). *Sustaining Global Surveillance and Response to Emerging Zoonotic Diseases*. Washington, DC: The National Academies Press.

85. Islam, M. K., Alim, M. A., Tsuji, N., and Mondal, M. M. H. (2006). An investigation into the distribution, host-preference and population density of Ixodid ticks affecting domestic animals in Bangladesh. *Trop. Anim. Health Prod.* 38, 485–490.

86. Jones, K. E., Patel, N. G., Levy, M. A., Storeygard, A., Balk, D., Gittleman, J. L., and Daszak, P. (2008). Global trends in emerging infectious diseases. *Nature* 451, 990–993.

87. Kistner, T. P., and Hayes, F. A. (1970). White-tailed deer as hosts of cattle fever ticks. *J. Wildl. Dis.* 6, 437–440.

88. Kjelland, M. E., Kreuter, U. P., Clendenin, G. A., Wilkins, R. N., Wu, X. B., Afanador, E. G., and Grant, W. E. (2007). Factors related to spatial patterns of rural land fragmentation in Texas. *Environ. Manage.* 40, 231–244.

89. Kuttler, K. L., Graham, O. H., Johnson, S. R., and Trevino, J. L. (1972). Unsuccessful attempts to establish cattle Babesia infections in white-tailed deer. *J. Wildl. Dis.* 8, 63–66.

90. Lafferty, K. (2009). The ecology of climate change and infectious diseases. *Ecology* 90, 888–900.

91. Lean, J. L. (2010). Cycles and trends in solar irradiance and climate. *Wiley Interdiscip. Rev. Clim. Change* 1, 111–122.

92. Leslie, D. M. (2008). *Boselaphus tragocamelus* (Artiodactyla: Bovidae). *Mamm. Species* 813, 1–16.

93. Linthicum, K. J., Britch, S. C., Anyamba, A., Small, J., Tucker, C. J., Chretien, J., and Sithiprasasna, R. (2008). "Ecology of Disease: the intersection of human and animal health," in *Vector-Borne Diseases: Understanding the Environmental, Human health, and Ecological Connections*. Washington, D.C.: The National Academies Press.

94. Logue, J. N. (1995). *Beyond the Germ Theory – The Story of Dr. Cooper Curtice*. College Station: Texas A&M University Press.

95. Lohmeyer, K. H., Pound, J. M., May, M. A., Kammlah, D. M., and Davey, R. B. (2011). Distribution of *Rhipicephalus*(*Boophilus*) *annulatus* (Acari: Ixodidae) infestations detected in the United States along the Texas/Mexico border.*J. Med. Entomol.* 48, 770–774.

96. Malone, J. B. (1989). Texas fever, two-headed calves and the Hatch Act – 100 years and counting for veterinary parasitology in the United States. *Vet. Parasitol.* 33, 3–29.

97. Martinez, A., and Hewitt, D. (2001). Sobrepoblacion de venado cola blanca en el noreste de Mexico. *Ciencia UANL*4, 177–181.

98. McBride, W. D., and Mathews, K. Jr. (2011). *The Diverse Structure and Organization of U.S. Beef Cow-Calf Farms*. EIB-73. Washington: U.S. Department of Agriculture, Economic Research Service.

99. McCulloch, R. N., and Lewis, I. J. (1968). Ecological studies of the cattle tick, *Boophilus microplus*, in the North Coast district of New South Wales. *Aust. J. Agric. Res.* 19, 689–710.

100. Miller, R. J., Almazan, G. C., Estrada, O. M., Davey, R. B., and George, J. E. (2008). "A survey for fipronil-and-ivermectin-resistant *Rhipicephalus* (*Boophilus*) *microplus* collected in northern Mexico and the options for the management of acaricide-resistant ticks with pesticides," *Paper Presented at VI Seminario Internacional de Parasitologia: Impacto de las enfermedades parasitarias sobre*

la ganaderia globalizad. INIFAP-INFARVET AMPAVE-CNG-UV. Boca del Rio, Veracruz.

101. Miller, R. J., Davey, R. B., and George, J. E. (2005). First report of organophosphate-resistant *Boophilus microplus*(Acari: Ixodidae) within the United States. *J. Med. Entomol.* 42, 912–917.
102. Miller, R. J., Davey, R. B., and George, J. E. (2007). First report of permethrin-resistant *Boophilus microplus*(Acari: Ixodidae) collected within the United States. *J. Med. Entomol.* 44, 308–315.
103. Miller, R. J., Estrada-Peña, A., Almazán, C., Allen, A., Jory, L., Yeater, K., Messenger, M., Ellis, D., and Pérez de León, A. A. (2012). Exploring the use of an anti-tick vaccine as a tool for the integrated eradication of the cattle fever tick, *Rhipicephalus (Boophilus) annulatus. Vaccine,* accepted.
104. Morgan, E. R., Milner-Gulland, E. J., Torgerson, P. R., and Medley, G. F. (2004). Ruminating on complexity: macroparasites of wildlife and livestock. *Trends Ecol. Evol.* 19, 181–188.
105. Mount, G. A., Haile, D. G., Davey, R. B., and Cooksey, L. M. (1991). Computer simulation of *Boophilus* cattle tick (Acari: Ixodidae) population dynamics. *J. Med. Entomol.* 28, 223–240.
106. Munderloh, U. G., and Kurtti, T. J. (2011). "Emerging and re-emerging tick-borne diseases: new challenges at the interface of human and animal health," in *National Academy of Sciences, Washington, D.C., Critical needs and gaps in understanding prevention, amelioration, and resolution of Lyme and other tick-borne diseases: the short-term and long-term outcomes,* A142–A166.
107. Mungall, E. C., and Sheffield, W. J. (1994). *Exotics on the Range.* College Station: Texas A&M University Press.
108. Munroe, F., and Willis, N. (2007). The role of anticipation in enhancing prevention and preparedness. *Vet. Ital.* 43, 207–214.
109. Munroe, F. A. (2007). Integrated agricultural intelligence – a proposed framework. *Vet. Ital.* 43, 215–223.
110. National Climatic Data Center. (2012). *2011 National Highlights.* Asheville, NC.
111. National Research Council of the National Academies. (2010). *Toward Sustainable Agricultural Systems in the 21st Century.* Washington, D.C.: The National Academies Press.

112. Newell, W., and Daugherty, M. S. (1906). *The Cattle Tick (Boophilus annulatus): Studies of the Egg and Seed Tick Stages, a Simple Method of Eradicating the Tick*. Circular No. 10. Baton Rouge, LA: State Crop Pest Commission of Louisiana.

113. Oberg, A. L., Kennedy, R. B., Li, P., Ovsyannikova, I. G., and Poland, G. A. (2011). Systems biology approaches to new vaccine development. *Curr. Opin. Immunol.* 23, 436–443.

114. Olson, S. H., and Patz, J. A. (2011). Global environmental change and tick-borne disease incidence," in *National Academy of Sciences, Washington, D.C., Critical needs and gaps in understanding prevention, amelioration, and resolution of Lyme and other tick-borne diseases: the short-term and long-term outcomes,* A44–A66.

115. Palmer, W. A., Treverrow, N. L., and O'Neill, G. H. (1976). Factors affecting the detection of infestations of*Boophilus microplus* in tick control programs. *Aust. Vet. J.* 52, 321–324.

116. Patarroyo, J. H., Portela, R. W., De Castro, R. O., Pimentel, J. C., Guzman, F., Patarroyo, M. E, Vargas, M. I., Prates, A. A., and Mendes, M. A. (2002). Immunization of cattle with synthetic peptides derived from the *Boophilus microplus* gut protein (Bm86). *Vet. Immunol. Immunopathol.* 88, 163–172.

117. Peel, M. C., Finlayson, B. L., and McMahon, T. A. (2007). Updated world map of the Koppen-Geiger climate classification. *Hydrol. Earth Syst. Sci.* 11, 1633–1644.

118. Pegram, R. G., Wilson, D. D., and Hansen, J. W. (2000). Past and present national tick control programs. Why they succeed or fail. *Ann. N. Y. Acad. Sci.* 916, 546–554.

119. Pérez de León, A. A., Strickman, D. A., Knowles, D. P., Fish, D., Thacker, E. L., De La Fuente, J., Krause, P. J., Wikel, S. K., Miller, R., Wagner, G. G., Almazan, C., Hillman, R., Messenger, M. T., Ugstad, P. O., Duhaime, R. A., Teel, P. D., Ortega-Santos, A., Hewitt, D. G., Bowers, E. J., Bent, S. J., Cochran, M. H., McElwain, T. F., Scoles, G. A., Suarez, C. E., Davey, R. B., Freeman, J. M., Lohmeyer, K. H., Li, A. Y., Guerrero, F., Kammlah, D. M., Phillips, P. L., and Pound, J. M. (2010). One Health approach to identify research needs in bovine and human babesioses: workshop report.*Parasit Vectors* 3, 36–46.

120. Perry, B., Grace, D., and Sones, K. (2011). Current drivers and future directions of global livestock disease dynamics. *Proc. Natl. Acad. Sci. U.S.A.* doi: 10.1073/pnas.1012953108. [Epub ahead of print].

121. Pound, J. M., George, J. E., Kammlah, D. M., Lohmeyer, K. H., and Davey, R. B. (2010). Evidence for the role of white-tailed deer (Artiodactyla:Cervidae) in the epidemiology of cattle ticks and southern cattle ticks (Acari:Ixodidae) in reinfestations along the Texas/Mexico border in South Texas – A review and update. *J. Econ. Entomol.* 103, 211–218.

122. Pound, J. M., Miller, J. A., George, J. E., and Fish, D. (2009). The United States Department of Agriculture Northeast Area-wide Tick Control Project: history and protocol. *Vector Borne Zoonotic Dis.* 9, 365–370.

123. Pretty, J., Sutherland, W., Ashby, J., Auburn, J., Baulcombe, D., Bell, M., Bentley, J., Bickersteth, S., Brown, K., Burke, J., Campbell, H., Chen, K., Crowley, E., Crute, I., Dobbelaere, D., Edwards-Jones, G., Funes-Monzote, F., Godfray, C. J., Griffon, M., Gypmantisiri, P., Haddad, L., Halavatau, S., Herren, H., Holderness, M., Izac, A., Jones, M., Koohafkan, P., Lal, R., Lang, T., McNeely, J., Mueller, A., Nisbett, N., Noble, A., Pingali, P., Pinto, Y., Rabbinge, R., Ravindranath, N., Rola, A., Roling, N., Sage, C., Settle, W., Sha, J., Shiming, L., Simons, T., Smith, P., Strzepeck, K., Swaine, H., Terry, E., Tomich, T., Toulmin, C., Trigo, E., Twomlow, S., Vis, J., Wilson, J., and Pilgrim, S. (2010). The top 100 questions of importance to the future of global agriculture. *Int. J. Agric. Sustain.* 8, 219–236.

124. Racelis, A. E., Davey, R. B., Goolsby, J. A., Pérez de León, A. A., Varner, K., and Duhaime, R. (2012). Facilitative ecological interactions between invasive species: *Arundo donax* stands as favorable habitat for cattle ticks (Acari: Ixodidae) along the U.S.-Mexico border. *J. Med. Entomol.* 49, 410–417.

125. Randolph, S. E. (2010). To what extent has climate change contributed to the recent epidemiology of tick-borne diseases? *Vet. Parasitol.* 167, 92–94.

126. Rawlins, S. C. (1979). Seasonal variation in the population density of larvae of *Boophilus microplus* (Canestrini) (Acari: Ixodidae) in Jamaican pastures. *Bull. Entomol. Res.* 69, 87–91.

127. Reisen, W. (2010). Landscape epidemiology of vector-borne diseases. *Annu. Rev. Entomol.* 55, 461–483.

128. Rollins, D., Higginbotham, B. J., Cearley, K. A., and Wilkins, R. N. (2007). Appreciating feral hogs: extension education for diverse stakeholders in Texas. *Hum. Wildl. Confl.* 1, 192–198.

129. Rosario-Cruz, R., Almazán-García, C., Miller, R. J., Domínguez-García, D. I., Hernandez-Ortiz, R., and de la Fuente, J. (2009). Genetic basis and impact of tick acaricides resistance. *Front. Biosci.* 14, 2657–2665.

130. Rubel, F., and Kottek, M. (2010). Observed and projected climate shifts 1901–2100 depicted by world maps of the Köppen-Geiger climate classification. *Meteorol. Z.* 19, 135–141.

131. Rue, L. L. (1978). *The Deer of North America*. New York: Crown Publishers, Inc.

132. Sanders, D. (2011). *Ticks and Tick-Borne Pathogens Associated with Feral Swine in Edwards Plateau and Gulf Prairies and Marshes Ecoregions of Texas*. Ph.D. dissertation, Texas A&M University, College Station.

133. Say, T. (1821). An account of the Arachnides of the United States. *J. Acad. Nat. Sci. Phila.* 2, 59–82.

134. Schultz, M. (2008). Theobald Smith. *Emerging Infect. Dis.* 14, 1940–1942.

135. Schuster, A. (2011). *Spatial and Temporal Survey of Feral Pig Ectoparasites in Three Texas Wildlife Districts*. Doctoral dissertation, Texas A&M University.

136. Scifres, C. J. (1980). *Brush Management: Principles and Practices for Texas and the Southwest*. College Station: Texas A&M University Press.

137. Scifres, C. J. (1985). "IBMS: Ecological Basis and Evolution of Concepts," in *Integrated Brush Management Systems for South Texas: Development and Implementation*. Publication No. 2M-6-85. College Station, TX: Texas Agricultural Experiment Station.

138. Scifres, C. J., and Hamilton, W. T. (1985). "Selecting IBMS Components," in *Integrated Brush Management Systems for South Texas: Development and Implementation*. Publication No. 2M-6-85. College Station, TX: Texas Agricultural Experiment Station.

139. SeaSolve. (2004). *AutoSignal: Pioneering Automated Signal Analysis and Advanced Filtering. Users Manual.* Framingham, MA: SeaSolve Software Inc.

140. Sheffield, W. J., Ables, E. D., and Fall, B. A. (1971). Geographic and ecologic distribution of nilgai antelope in Texas.*J. Wildl.* 35, 250–257.

141. Sheffield, W. J., Fall, B. A., and Bennet, A. B. (1983). *The Nilgai Antelope in Texas. Kleberg Studies in Natural Resources.* Publication No. 2M-10-83. College Station, TX: Texas Agricultural Experiment Station.

142. Simmons, J. (2011). *The Three Rights: Food, Choice, Sustainability – Technology's Role in the 21st Century: Making Safe, Affordable and Abundant Food a Global Reality.* Greenfield, IN: Elanco Animal Health.

143. Sitte, K., Brinkworth, R., East, I. J., and Jazwinska, E. C. (2002). A single amino acid deletion in the antigen binding site of BoLA-DRB3 is predicted to affect peptide binding. *Vet. Immunol. Immunopathol.* 85, 129–135.

144. Smith, J. E. (2007). Science and technology foresight: a provocative tool for contending with future challenges in food safety and public veterinary medicine. *Vet. Ital.* 43, 237–246.

145. Smith, T., and Kilbourne, F. L. (1893). *United States Department of Agriculture Bureau of Animal Industries Bulletin*, Washington: Government Printing Office.

146. Sonenshine, D. E., and Mather, T. N. (1994). *Ecological Dynamics of Tick-Borne Zoonoses.* New York: Oxford University Press.

147. Strom, C. (2009). *Making Catfish Bait out of Government Boys: The Fight against Cattle Ticks and the Transformation of the Yeoman South. (Environmental History and the American South).* Athens: University of Georgia Press.

148. Suk, J. E., Lyall, C., and Tait, J. (2008). Mapping the future dynamics of disease transmission: risk analysis in the United Kingdom Foresight programme on the detection and identification of infectious diseases. *Euro Surveill.*13,1–7.

149. Sutherst, R. (2001). The vulnerability of animal and human health to parasites under global change. *Int. J. Parasitol.* 31, 933–948.

150. Tabachnick, W. (2010). Challenges in predicting climate and environmental effects on vector-borne disease episystems in a changing world. *J. Exp. Biol.* 213, 946–954.

151. Taylor, R. B., Hellgren, E. C., Gabor, T. M., and Ilse, L. M. (1998). Reproduction of feral pigs in southern Texas. *J. Mammal.* 9, 1325–1331.

152. Taylor, W. P. (1956). *The Deer of North America*. Washington, D.C.: The Stackpole Company and The Wildlife Management Institute.

153. Teel, P. D. (1991). Application of modeling to the ecology of *Boophilus annulatus* (Say) (Acari: Ixodidae). *J. Agric. Entomol.* 8, 291–296.

154. Teel, P. D., Corson, M. S., Grant, W. E., and Longnecker, M. T. (2003). Simulating biophysical and human factors that affect detection probability of cattle fever ticks (Boophilus spp.) in semi-arid thornshrublands of South Texas.*Ecol. Modell.* 170, 29–43.

155. Teel, P. D., Marin, S. L., Grant, W. E., and Stuth, J. W. (1997). Simulation of host-parasite-landscape interactions: influence of season and habitat on cattle fever tick (Boophilus sp.) population dynamics in rotational grazing systems. *Ecol. Modell.* 97, 87–97.

156. Teel, P. D., Strey, O. F., and Hurley, J. A. (2011). The TickApp for Texas and the Southern Region: a mobile smart Phone application. Available at: http://tickapp.tamu.edu

157. Trapaga, J. T. (1989). *The Eradication of Ticks – La Campana Contra la Garrapata Boophilus spp en Mexico-Logros, Problemas y Perspectivas*. Rome: Food and Agriculture Organization of the United Nations, FAO, 114–121.

158. Vannier, E., Gewurtz, B., and Krause, P. J. (2008). Human babesiosis. *Infect. Dis. Clin. North Am.* 22, 469–488.

159. Vesco, U., Knap, N., Labruna, M. B., Avšič-Županc, T., Estrada-Peña, A., Guglielmone, A. A., Bechara, G. H., Gueye, A., Lakos, A., Grindatto, A., Conte, V., and De Meneghi, D. (2011). An integrated database on ticks and tick-borne zoonoses in the tropics and subtropics with special reference to developing and emerging countries. *Exp. Appl. Acarol.* 54, 65–83.

160. Walker, A. R. (2011). Eradication and control of livestock ticks: biological, economic and social perspectives.*Parasitology* 138, 945–959.

161. Walker, A. R., Bouattour, A., Camicas, J. L., Estrada Pena, A., Horak, I. G., Latif, A. A., Pegram, R. G., and Preston, P. M. (2003). *Ticks of Domestic Animals in Africa; A Guide to Identification of Species*. Edinburgh: Bioscience Reports.

162. Welburn, S. (2011). One health: the 21st century challenge. *Vet. Rec.* 168, 614–615.

163. Whalon, M. E., Mota-Sanchez, D., and Hollingsworth, R. M. (2008). *Global Pesticide Resistance in Arthropods*. Cambridge: CAB International.

164. Willadsen, P. (2008). "Anti-tick vaccines," in *Ticks: Biology, Disease and Control*, eds A. S. Bowman and P. A. Nuttall (Cambridge: Cambridge University Press), 424–446.

165. Willadsen, P., Bird, P., Cobon, G. S., and Hungerford, J. (1995). Commercialisation of a recombinant vaccine against*Boophilus microplus*. *Parasitology* 110, S43–S50.

166. Willis, N. G. (2007). The animal health foresight project. *Vet. Ital.* 43, 247–Woolhouse, M. (2011). How to make predictions about future infectious disease risks. *Philos. Trans. R. Soc. Lond. B Biol. Sci.* 366, 2045–2054.

167. Young, S. P. (1956). "The deer, the Indians, and the American pioneers," in *The Deer of North America*, ed. W. P. Taylor (Harrisburg, PA: The Stackpole Company), 1–28.

CHAPTER 2

Incorporating Within- And Between-Patch Resource Selection in Identification of Critical Habitat for Brood-Rearing Greater Sage-Grouse

Matthew R Dzialak[1], Chad V Olson[1], Stephen L Webb[1, 2], Seth M Harju[1], and Jeffrey B Winstead[1]

[1]Hayden-Wing Associates, LLC, Hatural Resource Consultants, 2308 South 8th Street, Laramie 82070, WY, USA

[2]The Samuel Roberts Noble Foundation, 2510 Sam Noble Parkway, Ardmore 73401, OK, USA

ABSTRACT

Introduction

Incorporating information on animal behavior in resource-based predictive modeling (e.g., occurrence mapping) can elucidate the relationship between process and spatial pattern and depict habitat in terms of its structure as well as its function. In this paper, we assigned location data on brood-rearing greater sage-grouse (*Centrocercus urophasianus*) to either within-patch (encamped) or between-patch (traveling) behavioral modes by estimating a movement-based relative displacement index. Objectives were to estimate and validate spatially explicit models of within- versus between-patch resource selection for application in habitat management and compare these models to a non-behaviorally adjusted model.

Results

A single model, the vegetation and water resources model, was most plausible for both the encamped and traveling modes, including the non-behaviorally adjusted model. When encamped, sage-grouse selected for taller shrubs, avoided bare ground, and were closer to mesic areas. Traveling sage-grouse selected for greater litter cover and herbaceous vegetation. Preference for proximity to mesic areas was common to both encamped and traveling modes and to the non-behaviorally adjusted model. The non-behaviorally adjusted map was similar to the encamped model and validated well. However, we observed different selection patterns during traveling that could have been masked had behavioral state not been accounted for.

Conclusions

Characterizing habitat that structured between-patch movement broadens our understanding of the habitat needs of brood-rearing sage-grouse, and the combined raster surface offers a reliable habitat management tool that is readily amenable to application by GIS users in efforts to focus sustainable landscape management.

INTRODUCTION

Greater sage-grouse (*Centrocercus urophasianus*; hereafter, sage-grouse) occurs throughout shrub-steppe habitat in 11 western American states and 2 Canadian provinces. The species is endangered in Canada (COSEWIC [2008]). The sage-grouse is not listed in the

US as federally threatened or endangered, although it was found to be warranted (United States Fish and Wildlife Service [2010]). Long-term population declines have been observed throughout much of the species distribution (Connelly and Braun [1997]; Connelly et al. [2004]; Fedy and Aldridge [2011]). Factors associated with decline include large-scale changes in temperature and precipitation cycles, fire, predation, and human activity associated with agriculture and energy development (Connelly and Braun [1997]; Connelly et al. [2000]; Walker et al. [2007]; Doherty et al. [2008]; Harju et al.[2010]; Holloran et al. [2010]).

Identifying resources necessary for critical life-history phases (e.g., brood-rearing, nesting, and roosting) and survival are important for managing landscapes (Dzialak et al. [2011b]), especially when landscapes are exposed to large-scale modification from climatic and anthropogenic sources. The needs of brooded hens and their chicks revolve around nutrition acquisition, protection from predation, water needs, and favorable thermal environments. Brood-rearing areas for sage-grouse include moderate shrub coverage with a prominent sagebrush component, proximity to mesic areas, and herbaceous vegetation such as forbs that also harbors insects (Hagen et al. [2007]; Dzialak et al. [2011b]; Harju et al. [2013a]). However, less is known about how behavior structures the use of resources during critical seasons.

Resource-based predictive modeling of species occurrence has strong application in wildlife management because it provides information on the relationship between animals and landscape features, and it offers spatially explicit guidance in identifying critical habitat throughout large areas (Johnson et al. [2004]; Guisan and Thuiller [2005]). One of the most useful and prevalent applications of species occurrence modeling is spatial prioritization of management activity, wherein habitat that is identified as having a high probability of importance to wildlife is managed differently than habitat that is predicted to be used with low frequency or otherwise of less importance (Margules and Pressey [2000]; Aldridge and Boyce [2007]; Sawyer et al. [2009]; Dzialak et al. [2011b]). Efforts to identify and prioritize important wildlife habitat are particularly relevant in human-modified areas where wildlife conservation may be one of the several valued land uses or where protecting large contiguous areas of unmodified habitat is not an option (Moilanen et al. [2005]). Thus, incorporating information on behavior in predictive modeling can play an important role in developing accurate predicted occurrence maps that depict habitat in terms of its composition and configuration for different behaviors (Taylor et al. [1993]; Tischendorf and Fahrig [2000]; Kindlmann and Burel [2008]; Van Moorter et al. [2010]).

Availability of spatial data has created opportunities to advance methods that link behavioral modes to processes such as population change or resource selection (Turchin [1998]; Forester et al. [2009]; Morales et al. [2010]). From a spatial perspective, a behavioral mode can be thought of as a manner of movement characterized by parameters such as movement rate, net displacement, or tortuosity (Patterson et al. [2008]). For example, in some species, behavior characterized by high directional persistence, a high rate of movement, and high displacement will reflect a between-patch mode such as traveling or migration, whereas high tortuosity, a low rate of movement, and low displacement will reflect a within-patch mode such as encampment, resting or foraging (Van Moorter et al. [2010]). From a management perspective, species-specific behavioral responses to the landscape establish a mechanistic basis for understanding how particular features function ecologically (Tischendorf and Fahrig [2000]). Behavioral mode is a variable that the investigator may wish to characterize before conducting further analysis (Van Moorter et al. [2010]) or incorporate as a latent process into structural equation or time-series models (Patterson et al. [2008]). Depending on the scale of investigation, dominant breeding season behavioral modes of sage-grouse might include pre-nesting behavior such as lek attendance, nest-site selection, nesting off-bout behavior, and between-versus within-patch movement of the brooded female (Aldridge and Brigham [2002]; Aldridge and Boyce [2007]; Doherty et al. [2010]; Dzialak et al. [2011b]). Global positioning system (GPS) data at a fine temporal scale make transitions between these modes readily observable (Dzialak et al. [2011b]). Identifying behavioral mode within the context of current management establishes the opportunity to quantitatively assign observed data to a defined mode as an initial analytical step. Incorporating datum-specific information on behavior into subsequent spatial modeling enables the investigator to depict the landscape in terms of its structure (composition and configuration of patches) relative to a given behavior mode, offering richer insight for habitat management (Taylor et al. [1993]; Kindlmann and Burel [2008]).

The objective of this work was to identify within-patch (encamped) versus between-patch (traveling) brood-rearing habitat for the sage-grouse throughout a portion of its Wyoming, USA, and distribution. Preliminary GPS data revealed a pattern observed across all sampled individuals involving movement between and within patches (Figure 1). We aimed to assign location data to encamped versus traveling behavioral modes based on movement parameters estimated from GPS data, estimate resource selection functions for these two respective modes, and compare the behavioral models to a non-behaviorally adjusted model (including all data not assigned to behavioral state). To provide spatial products for application in prioritization of management, we developed spatially explicit estimates of occurrence separately for encamped and traveling modes. We integrated encamped and traveling raster surfaces to depict (i.e., map) areas important during within-patch movement, between-patch movement, and concurrently in both behavioral modes. We validated predictions using location data withheld from model development.

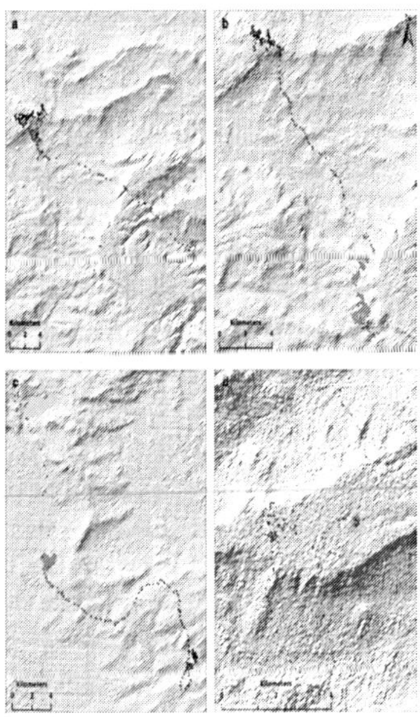

Figure 1: GPS locations and movement patterns of greater sage-grouse in Wyoming. All greater sage-grouse (*Centrocercus urophasianus*) comprising the sample showed within-patch (clusters of GPS locations) and between-patch (trails between clusters) movement in south-central Wyoming, USA,

during 2009 to 2012. Based on the relative displacement index, red dots indicate locations assigned to the encamped behavioral mode and black dots indicate locations assigned to the traveling mode; yellow triangles show the nest location. Panels (a-d) each show one unique individual for the years 2009 to 2012, respectively.

METHODS

Study Area

The 7,948-km^2 study area encompassed portions of the Great Divide Basin in south-central Wyoming, USA. Terrain is characterized by rolling sagebrush steppe interspersed with gently sloping flats, drainages, and vegetated sand dunes; elevation is 1,933 to 2,385 m. Dominant vegetation at lower elevation included Wyoming big sagebrush (*Artemisia tridentata wyomingensis*), birdfoot sagebrush (*A. pedatifida*), black greasewood (*Sarcobatus vermiculatus*), shadscale (*Atriplex confertifolia*), and grass species (e.g., *Pseudoroegneria spicata, Agropyron* spp., *Bromus* spp.). At higher elevation, mountain big sagebrush (*A. t. vaseyana*) was present; basin big sagebrush (*A. t. tridentata*) occurred in drainages. Average monthly precipitation during the study period was 2.68 cm (Western Regional Climate [2011]). Predominant historic and ongoing land uses were domestic livestock grazing and development of energy resources. As of March 2012, there were 4,894 active wells within the study area (1 well/1.62 km^2).

Field Procedures and Location Data

During 2008 to 2011, we captured female sage-grouse with nets and the aid of binoculars and spotlights (Wakkinen et al. [1992]) during spring (generally March and April) on and around leks that were dispersed throughout the study area. We determined age (Crunden [1963]) and fitted sage-grouse with either 22 or 30-g ARGOS/GPS Solar PTTs (PTT-100, Microwave Telemetry Inc., Columbia, MD, USA) using a rump-mount technique (Dzialak et al. [2012]). GPS units had a 3-year operational life and were configured with ultra-high frequency (UHF) beacons that enabled ground tracking for field-based confirmation of events (i.e., fatality). For this analysis, we used brood-rearing data defined uniquely for each bird; brood-rearing across birds occurred from 11 May to 12 August, 2009 to 2012. Collars were programmed to record location information every 1 h from 0700 to 2200 h. The use of GPS made the transition from nesting to brood-rearing readily observable, with the initiation of brood-rearing defined for each bird as the specific GPS location at which the female moved from the nest with no subsequent return to the nest site (Dzialak et al.[2011b]). Brood surveys were conducted weekly beginning when chicks were about 7 days old. An effort was made to determine presence versus absence of a brood without flushing females. The presence of chicks was determined by direct observation of chicks and based on behavior of the female (i.e., the female walking or running away from the observer rather than flying, becoming defensive or aggressive, or displaying wing-dragging or flutter-hopping behavior) (Patterson[1952]). Animal capture and handling protocols were approved by the Wyoming Game and Fish Department (Chapter 33 Permit #649).

Assigning Location Data to Behavioral Mode

We developed a relative displacement index to assign location data to encamped versus traveling behavioral modes. The index is based on net displacement or the straight-line distance between the starting location for an individual and each subsequent location along the movement path of that individual (Turchin [1998]). Net displacement was calculated for each location within each individual's movement path as:

$$\text{Net displacement}_l = \sqrt{\{(UTM_{el} - UTM_{e1})^2 + (UTM_{nl} - UTM_{n1})^2\}} \qquad (1)$$

Where UTMe and UTMn are respective easting and northing coordinates of location data (datum and projection were North American Datum of 1983 and Universal Transverse Mercator zone 13 north), 1 is the first GPS location in an individual's path, and *l* is each subsequent location in the individual's path from 1 to *l* locations. We calculated the relative displacement index as:

$$\left|\{(\text{net displacement}_l - \text{net displacement}_{l+1}) \div \text{net displacement}_l\} \times 100\right| \qquad (2)$$

which quantifies the relative change in amplitude of net displacement between location l and location $l+1$. We assigned a cut-off associated with the relative displacement index of 5% based on visual inspection and preliminary analysis of the location data. Locations considered as encamped did not vary substantially in amplitude, and the frequency of locations with a relative displacement index of ≤5% was much greater compared to the frequency of locations with a relative displacement index >5%; these locations also corresponded with much longer movement distances. Specifically, for any instance in which the relative change in amplitude of net displacement from location l to location $l+1$ was ≤ 5%, we assigned the behavioral mode as encamped; otherwise, we assigned the behavioral mode as traveling (relative displacement index > 5%; Figures 1 and 2).

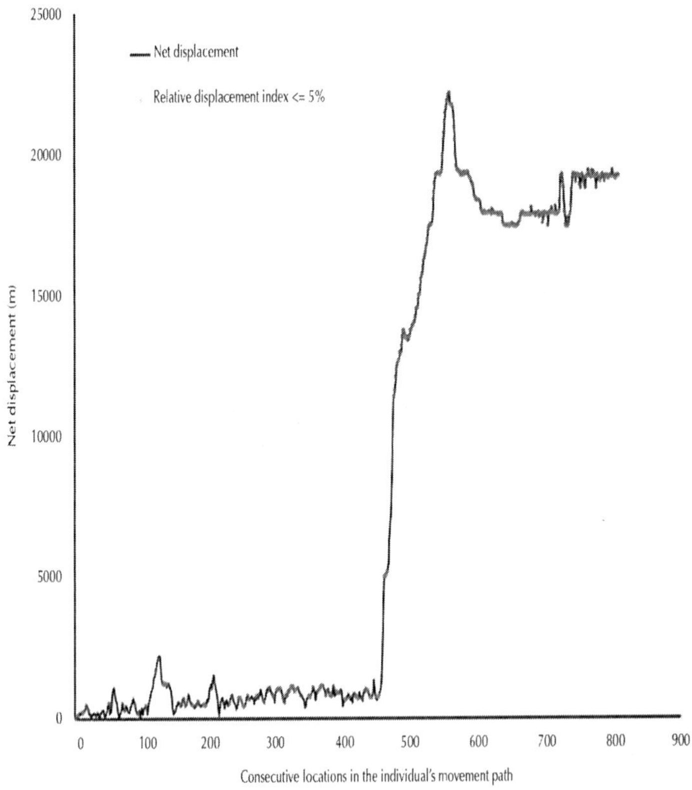

Figure 2: Relationship between net displacement and relative displacement index for assigning behavioral state to greater sage-grouse. General relation-

ship between net displacement and the relative displacement index for greater sage-grouse (*Centrocercus urophasianus*) in south-central Wyoming, USA, from 2009 to 2012. The black line illustrates net displacement or the straight-line distance between the starting location for an individual and subsequent locations along the movement path of that individual. Red dots highlight those locations for which the relative displacement index (%), or relative change in amplitude of net displacement between location *I* and location *I*+1, is ≤5%, which identifies the encamped behavioral mode.

Predictor Variables

Using a geographic information system (GIS; ArcGIS 10, ESRI, Redlands, CA, USA), we calculated predictor variables depicting landscape features that influence resource selection behavior of sage-grouse (Table 1). This suite of variables has provided a basis for accurate prediction of occurrence and demographic responses among sage-grouse as part of companion studies (Dzialak et al.[2011b]; [2012]; [2013a]; [2013b]; Webb et al. [2012]). Raster images of oil and natural gas wells were updated annually so that we could analyze sage-grouse location data against the most up-to-date information on the location of wells. Raster images of all other human modification of the landscape were developed based on 2006 aerial imagery and updated using 2009 imagery (Table 1). These images were developed via heads-up-digitizing, a process whereby linear and areal features displayed in remotely sensed imagery are identified, interpreted, delineated, and attributed manually in a GIS by an investigator and then converted to raster data. We used spatial analyst in ArcGIS to develop all raster images and to extract cell values from raster images to location data (i.e., used and non-used points) for all predictor variables. All raster data were calculated at a resolution of 30 m (Table 1). We natural log-transformed distance variables (after adding 0.1 to all values; Table 1) to allow a functional form of the relationship between resource selection and distance that depicted a decreasing magnitude of influence with increasing distance. We also developed quadratic terms for vegetation and topographic variables because sage-grouse have been shown to avoid the lowest and highest values associated with some landscape features (Aldridge and Boyce [2007]; Dzialak et al. [2011b]; [2012]) and to assess potential thresholds.

Table 1: Predictor variables important to resource selection by greater sage-grouse in Wyoming

Variable	Description
Predominant human modifications of the landscape	
Distance to nearest well	Distance (m) to the nearest oil or natural gas well. Data on the location of wells, as of March 2012, were obtained from the Wyoming Oil and Gas Conservation Commission. All distance covariates were calculated using the Spatial Analyst\Euclidean Distance tool in ArcMap
Distance to nearest road	Distance (m) to paved, improved un-paved, or maintained dirt roads. Two-track roads were not included. Roads were heads-up-digitized at a scale of 1:2,000 from 2006 and 2009 National Agriculture Imagery Program aerial imagery at 1-m resolution
Distance to nearest residential or agricultural structure	Distance (m) to the nearest residential or agricultural structure including houses, sheds, and barns. Structures were heads-up-digitized as described above
Distance to nearest energy-related ancillary feature	Distance (m) to the nearest infrastructure associated with energy development other than wells. Such ancillary features included compressor stations, settling ponds, and buildings. Ancillary features were heads-up-digitized as described above
Predominant vegetation in the study area	
Percent shrub	Estimated percent of each pixel comprising the raster surface for which the vegetation type is shrub species. The data source was the Provisional Remote Sensing Sagebrush Habitat Quantification Products for Wyoming developed by the US Geological Survey. Detailed information on development and accuracy of all vegetation layers is in (Homer et al. [2012])
Percent sagebrush	Estimated percent of each pixel comprising the raster surface for which the vegetation type is sagebrush (*Artemisia* spp.). See Homer et al. ([2012])
Percent herbaceous	Estimated percent of each pixel comprising the raster surface that is herbaceous cover. See Homer et al. ([2012])
Percent bare ground	Estimated percent of each pixel comprising the raster surface that is bare ground. See Homer et al. ([2012])

Percent litter	Estimated percent of each pixel comprising the raster surface that is herbaceous litter. See Homer et al. ([2012])
Estimated shrub height	Estimated pixel-wide height (cm) of shrub vegetation. See Homer et al. ([2012])
Topographic and other natural features of the landscape	
Convexity 90 m	Using a digital elevation model (DEM; 1 arc-second National Elevation Dataset [NED] re-sampled to 30 m; available athttp://nationalmap.gov/viewer.html), convexity was calculated as the DEM pixel value minus the average elevation within a 90 × 90 m moving window (3 × 3 pixels). The average value of elevation within the moving window was calculated using the Spatial Analyst\Local\Cell Statistics tool selecting mean as the overlay statistic
Heat load index (HLI)	Rescaling of aspect (; radians calculated from the NED using Spatial Analyst in GIS) from 0 to 1 oriented northeast to southwest depicting the gradient from coolest to warmest aspect using the equation of McCune and Keon ([2002]): HLI = 1 to cosine(– 45)/2
Slope	Steepness (degrees) calculated from the NED using Spatial Analyst
Terrain roughness 90 m	An index of terrain roughness calculated as the standard deviation (SD) of elevation from a DEM within a 90 × 90 m moving window (3 × 3 pixels). This covariate was calculated using the Spatial Analyst\Local\Cell Statistics tool selecting SD as the overlay statistic
Distance to nearest mesic area	Distance (m) to the nearest permanent or intermittent stream, seep, spring, impoundment, irrigation, or water discharge area. The raster image was developed from 0.3-m true-color and CIR aerial photography (2009) using Feature Analyst® 4.2 (Visual Learning Systems [2008]) for ArcGIS® 10

Predictor variables calculated in a Geographic Information System (GIS; Esri ArcMap10) for potential inclusion in models of resource selection by greater sage-grouse (*Centrocercus urophasianus*) tracked during 2009 to 2012 in south-central Wyoming, USA. A description of each variable is provided in the right-hand column. All data (raster images) were calculated at a resolution of 30 m.

Dzialak *et al.*

Dzialak *et al. Ecological Processes* 2015 4:5, doi: 10.1186/s13717-015-0032-2

Modeling Resource Selection

We estimated resource selection using conditional logistic regression within an information-theoretic framework (Burnham and Anderson [2004]). To identify potentially informative versus uninformative predictor variables (*sensu* Arnold [2010]), we conducted univariate conditional logistic regression using the PHREG procedure in SAS. For each variable, we assessed the difference in Akaike's information criterion (AIC) between the null model and each univariate model. If the univariate model was within 4 AIC units of the null model, then it was considered uninformative and was dropped from further consideration. We then checked for correlation among retained variables using the correlation procedure (PROC CORR) in SAS; if correlation was very high (Pearson product-moment correlation $\geq |0.8|$), we eliminated the variable that was thought to be less relevant biologically (Drut et al. [1994]; Sveum et al. [1998]; Connelly et al. [2000]; Thompson et al. [2006]). Retained variables were then structured into a candidate set of models that depicted 15 competing hypotheses on how landscape features structured occurrence (Table 2; Walker et al. [2007]; Doherty et al. [2008]; Dzialak et al. [2011b]); we also assessed a null and global model for a total of 17 candidate models (i.e., competing hypotheses).

Table 2: Candidate models and variables used to assess resource selection in greater sage-grouse

Model	**Parameters**[a]	**K**
Null	None[b]	0
Sagebrush	Percent sagebrush (linear and quadratic terms)	2
Water resources	Percent sagebrush (linear and quadratic terms), distance to the nearest mesic area, and convexity 90 m	4
Sagebrush and non-energy-related anthropogenic features	Percent sagebrush (linear and quadratic terms), distance to the nearest mesic area, and distance to nearest residential or agricultural structure	4

Sagebrush and energy-related anthropogenic features	Percent sagebrush (linear and quadratic terms), distance to the nearest road, distance to nearest energy-related ancillary feature, and distance to nearest well	5
Terrain/topography	Percent sagebrush (linear and quadratic terms), slope (linear and quadratic terms), terrain roughness 90 m, and convexity 90 m	6
Movement facilitation	Percent sagebrush (linear and quadratic terms), percent bare ground (linear and quadratic terms), terrain roughness 90 m, and distance to the nearest road	6
Foraging	Percent sagebrush (linear and quadratic terms), percent herbaceous (linear and quadratic terms), distance to the nearest mesic area, and distance to the nearest residential or agricultural structure	6
Agriculture	Percent sagebrush (linear and quadratic terms), percent litter (linear and quadratic terms), distance to the nearest mesic area, and distance to the nearest residential or agricultural structure	6
Thermal environment	Percent sagebrush (linear and quadratic terms), percent shrub (linear and quadratic terms), slope (linear and quadratic terms), and heat load index	7
Local concealment	Percent sagebrush (linear and quadratic terms), percent shrub (linear and quadratic terms), percent bare ground (linear and quadratic terms), and percent litter (linear and quadratic terms)	8
Terrain/topography and foraging	Percent sagebrush (linear and quadratic terms), percent herbaceous (linear and quadratic terms), distance to the nearest mesic area, and distance to the nearest residential or agricultural structure, slope (linear and quadratic terms), terrain roughness 90 m, and convexity 90 m	10
Vegetation	Percent sagebrush (linear and quadratic terms), percent shrub (linear and quadratic terms), percent herbaceous (linear and quadratic terms), percent litter (linear and quadratic terms), percent bare ground (linear and quadratic terms), and estimated shrub height (linear and quadratic terms)	12

Foraging/perceived risk	Percent sagebrush (linear and quadratic terms), percent herbaceous (linear and quadratic terms), percent bare ground (linear and quadratic terms), estimated shrub height (linear and quadratic terms), distance to the nearest mesic area, distance to the nearest road, distance to the nearest energy-related ancillary feature, and distance to the nearest well	13
Vegetation and water resources	Percent sagebrush (linear and quadratic terms), percent shrub (linear and quadratic terms), percent herbaceous (linear and quadratic terms), percent litter (linear and quadratic terms), percent bare ground (linear and quadratic terms), estimated shrub height (linear and quadratic terms), and distance to the nearest mesic area	13
Vegetation and terrain/topography	Percent sagebrush (linear and quadratic terms), percent shrub (linear and quadratic terms), percent herbaceous (linear and quadratic terms), percent litter (linear and quadratic terms), percent bare ground (linear and quadratic terms), estimated shrub height (linear and quadratic terms), slope (linear and quadratic terms), terrain roughness 90 m, and convexity 90 m	16
Global	All parameters	22

[a]Listed in Table 1. [b]The PHREG procedure in SAS does not estimate an intercept. Candidate model set that includes competing hypotheses (null model, global model, and 15 competing hypotheses) on how landscape features structure occurrence of greater sage-grouse (*Centrocercus urophasianus*) during 2009 to 2012 in south-central Wyoming, USA. Each model (i.e., competing hypothesis) lists the number of parameters (*K*) and their description. Models are sorted in ascending order from least to most parameters.

Dzialak *et al.*

Dzialak *et al.* *Ecological Processes* 2015 **4**:5, doi: 10.1186/s13717-015-0032-2

To define resource availability, we calculated the distance (*D*) between successive grouse locations l_t and l_{t+1}. The spatial domain from which alternative choices to l_{t+1} (non-used locations) could be drawn was defined as a circular buffer centered on l_t with a radius equal to *D* between l_t and l_{t+1} plus 20% of that distance. We enforced a minimum buffer distance (i.e., 56 m) that was equal to the median distance between successive locations (i.e., *D*) observed across all individuals to acknowledge there was a minimum area available even if an individual did not move during that time period on a given day. We matched each used location with a set of five non-used locations drawn from within the circular buffer and considered each used location with its associated five non-used locations as a single stratum. This discrete-choice design (Cooper and Millspaugh [1999]; Compton et al. [2002]; McDonald et al. [2006]) quantifies a choice made by an individual female sage-grouse (i.e., used location) relative to five alternative choices that also were available temporally and spatially but were not chosen (i.e., non-used locations). One benefit of discrete-choice models is that inference is conditional on individual strata (e.g., a single used location and the paired non-used locations), thus accounting for potential spatiotemporal or within-individual autocorrelation among used locations (Pendergast et al. [1996]; Johnson et al. [2004]; Baasch et al. [2010]; Cushman and Lewis [2010]). We estimated resource selection models across all individuals and years to make population-level inference using conditional logistic regression for encamped and traveling behavioral modes separately, and then combined all behavior-specific data back into a non-behaviorally adjusted model for comparison. For the non-behaviorally adjusted model, we included a weight statement to adjust for the greater number of locations assigned to the encamped state. Locations assigned to the encamped mode were given a weight of 0.5138 whereas the traveling locations were given a weight of 1.0. AIC adjusted for small sample size difference (ΔAIC_c) and Akaike weights (*w*) were used to assess and select the most parsimonious models (Burnham and Anderson [2004]) of occurrence during encamped and traveling modes and for the non-behaviorally adjusted model.

Mapping the Response and Model Validation

Using the raster calculator tool in Spatial Analyst, we created natural log-transformed grids for distance variables, as well as grids depicting a quadratic form for terrain and vegetation variables. Models were made spatial using the raster calculator tool to invoke the equation:

$$w(x) = \exp\left(\sum_{K-1}^{K} \beta_k^x x_k\right) \qquad (3)$$

Where $w(x)$ is the relative probability of use, β_k^x is the parameter estimate for covariate k ($k = 1 \ldots K$), and raster data have values x (Manly et al. [2002]). Separate raster surfaces were generated for predicted occurrence during encamped and traveling behavioral modes and for the non-behaviorally adjusted model. Using SAS (rank and means procedures) and GIS, we partitioned cells comprising raster surfaces into deciles based on cell value, resulting in ten ranks for the relative probability of occurrence (1 = lowest, 10 = highest).

We withheld seven sage-grouse (20% of the total sample) from the model development as an independent validation sample; the validation sample was selected using a random number generator. We validated encamped and traveling raster surfaces by plotting 2,752 and 1,804 validation locations from the withheld sage-grouse on each surface, respectively, and testing whether the number of locations increased monotonically with bin rank (bins 1 to 10) using Spearman rank correlation ( ) implemented by the CORR procedure in SAS. We used another independent set of sage-grouse ($n = 7$) with 4,174 locations to validate the non-behaviorally adjusted raster surface. To generate a final map of predicted occurrence, we combined encamped and traveling raster layers using Spatial Analyst\Local\Combine, which provided an output raster in which a unique output value was assigned to each unique combination of input values. We reclassified the combined surface into the occurrence probability classes: high-priority encamped habitat; high-priority traveling habitat; high-priority encamped and traveling habitat; and low priority habitat.

RESULTS

Fix success rate during the brood-rearing season was high, averaging 94.0% (±1.7% SD). For model development, we used 19,557 GPS locations across 28 female sage-grouse; all locations were used for estimating resource selection of the non-behaviorally adjusted model. The relative displacement index assigned 12,919 locations to the encamped behavioral mode, and 6,638 locations to the traveling behavioral mode. Assignment to behavioral mode provided the basis for modeling resource selection during encamped and traveling modes (*sensu* Figures 1 and 2).

A single model was most plausible for each of the behavioral modes (Table 3), as well as for the combined, non-behaviorally adjusted model (Table 4). For both encamped and traveling modes and the non-behaviorally adjusted model, there was strong evidence in favor of the hypothesis that vegetation and water resources underpinned occurrence patterns of brooded females compared to all other hypotheses tested (Table 4). The vegetation and water resources model for encamped had 95% of model support and was 5.75 AIC_c units better than the next best model, which only had a model weight (w) of 0.05 (Table 3). During the traveling mode, the vegetation and water resources model had overwhelming model support ($w = 0.97$) compared to the next best model ($w = 0.03$; $AIC_c = 7.22$) (Table 3). Compared to the behavioral models, the vegetation and water resources model for the non-behaviorally adjusted data (Table 4) still had strong support ($w = 0.88$) but not as much support as the individual behavioral models.

Table 3: Ranking of candidate models for encamped and traveling modes of greater sage-grouse

Modes	***K***	**AIC_c**		***w***
Encamped				
Model				
Vegetation and water resources	13	33,663.78	0.00	0.95
Global	22	33,669.52	5.75	0.05
Foraging/perceived risk	13	33,769.51	105.73	0.00
Agriculture	6	33,835.05	171.28	0.00
Vegetation	12	33,895.72	231.95	0.00
Vegetation and terrain/topography	16	33,901.36	237.58	0.00
Local concealment	8	33,987.29	323.51	0.00
Sagebrush and non-energy-related anthropogenic features	4	33,988.50	324.72	0.00
Water resources	4	33,989.64	325.86	0.00
Foraging	6	33,990.89	327.11	0.00
Terrain/topography and foraging	10	33,994.76	330.99	0.00
Movement facilitation	6	34,050.88	387.10	0.00

Thermal environment	7	34,244.93	581.15	0.00
Sagebrush and energy-related anthropogenic features	5	34,321.29	657.52	0.00
Sagebrush	2	34,328.50	664.72	0.00
Terrain/topography	6	34,330.78	667.00	0.00
Null	0	34,671.67	1 007.90	0.00
Traveling				
Model				
Vegetation and water resources	13	17,540.62	0.00	0.97
Vegetation	12	17,547.84	7.22	0.03
Global	22	17,552.09	11.48	0.00
Vegetation and terrain/topography	16	17,554.23	13.61	0.00
Agriculture	6	17,557.83	17.21	0.00
Local concealment	8	17,562.81	22.20	0.00
Foraging/perceived risk	13	17,575.16	34.54	0.00
Movement facilitation	6	17,582.81	42.20	0.00
Foraging	6	17,605.54	64.93	0.00
Thermal environment	7	17,607.75	67.13	0.00
Sagebrush and non-energy-related anthropogenic features	4	17,609.30	68.69	0.00
Terrain/topography and foraging	10	17,610.45	69.83	0.00
Water resources	4	17,611.45	70.84	0.00
Sagebrush	2	17,624.95	84.34	0.00
Sagebrush and energy-related anthropogenic features	5	17,625.16	84.55	0.00
Terrain/topography	6	17,628.74	88.13	0.00
Null	0	17,733.84	193.23	0.00

List of candidate models (i.e., competing hypotheses) for encamped and traveling modes of greater sage-grouse (*Centrocercus urophasianus*) in south-central Wyoming, USA, from 2009 to 2012. Models are ranked in order of most plausible, including number of parameters (*K*), Akiake information scores corrected for small sample size (AIC_c), difference in AIC_c scores (Δ), and AIC_c weights (*w*).

Dzialak *et al.*

Dzialak *et al.* *Ecological Processes* 2015 4:5, doi: 10.1186/s13717-015-0032-2

Table 4: Ranking of candidate models for the non-behaviorally adjusted model of greater sage-grouse

Model	*K*	AIC_c		*w*
Vegetation and water resources	13	28,266.89	0.00	0.88
Foraging/perceived risk	13	28,270.86	3.97	0.12
Global	22	28,279.22	12.33	0.00
Agriculture	6	28,362.19	95.31	0.00
Vegetation	12	28,384.96	118.07	0.00
Vegetation and terrain/topography	16	28,391.44	124.55	0.00
Local concealment	8	28,446.15	179.26	0.00
Movement facilitation	6	28,480.97	214.08	0.00
Sagebrush and non-energy-related anthropogenic features	4	28,483.76	216.88	0.00
Foraging	6	28,484.18	217.29	0.00
Water resources	4	28,484.72	217.83	0.00
Terrain/topography and foraging	10	28,488.96	222.07	0.00
Thermal environment	7	28,593.09	326.20	0.00
Sagebrush and energy-related anthropogenic features	5	28,656.49	389.60	0.00
Sagebrush	2	28,659.07	392.18	0.00
Terrain/topography	6	28,662.48	395.59	0.00
Null	0	28,928.05	661.17	0.00

List of candidate models (i.e., competing hypotheses) for the non-behaviorally adjusted model of resource selection for greater sage-grouse (*Centrocercus urophasianus*) in south-central Wyoming, USA, from 2009 to 2012. Models are ranked in order of most plausible, including number of parameters (*K*), Akiake information scores corrected for small sample size (AIC_c), difference in AIC_c scores (Δ), and AIC_c weights (*w*).

Dzialak *et al.*

Dzialak *et al. Ecological Processes* 2015 4:5, doi: 10.1186/s13717-015-0032-2

When encamped, sage-grouse selected for taller shrubs and proximity to mesic areas, but avoided bare ground (Table 5). When traveling, sage-grouse selected for greater litter cover, herbaceous vegetation, and proximity to mesic areas. Sage-grouse during the encamped mode showed an affinity to greater shrub coverage, including sagebrush; during traveling, sage-grouse preferred taller shrubs but at a lower percentage of shrub coverage (Table 5). Resource selection patterns of sage-grouse in the non-behaviorally adjusted model had the same patterns of selection as the encamped model, albeit the magnitude of coefficient estimates was different (Table 6).

Table 5: Covariate parameter estimates of the most plausible model during encamped and traveling modes for greater sage-grouse

Modes	Estimate	SE	**95% CL**
Encamped			
Parameter			
Percent sagebrush (linear)	0.04	0.05	−0.06 to 0.15
Percent sagebrush (quadratic)	−0.003	0.003	−0.009 to 0.003
Percent shrub (linear)	0.13	0.07	−0.01 to 0.26
Percent shrub (quadratic)	−0.004	0.003	−0.01 to 0.001
Estimated shrub height (linear)	0.04	0.01	*0.02 to 0.07*
Estimated shrub height (quadratic)	−0.0004	0.001	−0.001 to 0.0002
Percent bare ground (linear)	−0.05	0.02	*−0.09 to −0.02*
Percent bare ground (quadratic)	0.0003	0.0002	*1.8 E−5 to 0.0007*
Percent litter (linear)	0.003	0.02	−0.03 to 0.04
Percent litter (quadratic)	−0.0001	0.0004	−0.0009 to 0.0008
Percent herbaceous (linear)	0.02	0.02	−0.02 to 0.06
Percent herbaceous (quadratic)	−0.0003	0.0007	−0.002 to 0.001
Distance to nearest mesic area	−0.10	0.01	*−0.13 to −0.08*
Traveling			
Parameter			
Percent sagebrush (linear)	−0.001	0.04	−0.09 to 0.09
Percent sagebrush (quadratic)	0.0001	0.002	−0.004 to 0.005
Percent shrub (linear)	−0.06	0.06	−0.18 to 0.07

Percent shrub (quadratic)	0.002	0.002	−0.003 to 0.007
Estimated shrub height (linear)	0.01	0.009	−0.005 to 0.03
Estimated shrub height (quadratic)	5.0 E−5	0.0002	−0.0003 to 0.0004
Percent bare ground (linear)	−0.02	0.02	−0.06 to 0.02
Percent bare ground (quadratic)	0.0001	0.0002	−0.0002 to 0.0005
Percent litter (linear)	0.07	0.02	*0.03 to 0.10*
Percent litter (quadratic)	−0.001	0.0005	*−0.002 to −0.0004*
Percent herbaceous (linear)	0.007	0.02	−0.04 to 0.05
Percent herbaceous (quadratic)	2.0 E−5	0.0008	−0.002 to 0.002
Distance to nearest mesic area	−0.04	0.02	*−0.08 to −0.005*

Coefficient estimates, standard error (SE), and 95% confidence limits (CL) for covariates from the most plausible model (vegetation and water resources; see Table 3) of resource selection by greater sage-grouse (*Centrocercus urophasianus*) during encamped and traveling behavioral modes in south-central Wyoming, USA from 2009 to 2012. Confidence limits in italics did not include zero.

Dzialak *et al.*

Dzialak *et al. Ecological Processes* 2015 4:5, doi: 10.1186/s13717-015-0032-2

Table 6: Covariate parameter estimates of the most plausible model of the non-behaviorally adjusted model for greater sage-grouse

Parameter	**Estimate**	**SE**	**95% CL**
Percent sagebrush (linear)	0.03	0.038	−0.04 to 0.11
Percent sagebrush (quadratic)	−0.002	0.002	−0.01 to 2.5 E−3
Percent shrub (linear)	0.053	0.057	−0.06 to 0.16
Percent shrub (quadratic)	−0.002	0.002	−0.01 to 2.2 E−3
Estimated shrub height (linear)	0.029	0.009	*0.01 to 0.05*

Estimated shrub height (quadratic)	2.0 E−4	2.0 E−4	−6.0 E−4 to 2.0 E−4
Percent bare ground (linear)	−0.037	0.014	*−0.06 to −0.01*
Percent bare ground (quadratic)	3.0 E−4	1.0 E−4	*1.0 E−5 to 5.0 E−4*
Percent litter (linear)	0.029	0.012	*4.6 E−3 to 0.05*
Percent litter (quadratic)	−0.001	3.0^{e-4}	−1.2 E−3 to 1.0 E−4
Percent herbaceous (linear)	0.013	0.014	−0.01 to 0.04
Percent herbaceous (quadratic)	3.0 E−4	0.001	−1.4 E−3 to 8.0 E−4
Distance to nearest mesic area	−0.084	0.017	*−0.12 to −0.05*

Coefficient estimates, standard error (SE), and 95% confidence limits (CL) for covariates from the non-behaviorally adjusted model (vegetation and water resources; see Table 4) of resource selection by greater sage-grouse (*Centrocercus urophasianus*) in south-central Wyoming, USA, from 2009 to 2012. Confidence limits in italics did not include zero.

Dzialak *et al.*

Dzialak *et al. Ecological Processes* 2015 4:5, doi: 10.1186/s13717-015-0032-2

The map depicting the encamped behavioral mode validated well (=0.99, $df=9$, $P<0.001$) with 9.3% of validation locations occurring in the lowest four bins and 70.6% in the highest four bins. Likewise, the map depicting the traveling behavioral mode validated well (=0.99, $df=9$, $P<0.001$) with 21.8% of validation locations occurring in the lowest four bins and 61.3% in the highest four bins. The non-behaviorally adjusted map validated well (=0.988, $df=9$, $P<0.001$); 69.5% of locations were in the highest four bins and 13.1% were in the lowest four bins. The final map in which encamped and traveling raster surfaces were combined depicts high-priority habitat for brood-rearing sage-grouse (Table 7, Figure 3).

Table 7: Reclassification of relative probability bins into broad designations that prioritized habitat based on behavioral mode

Bin rank				
Encamped	**Traveling**	**Designation**	**Number of cells**	**Percentage of landscape**
1 to 7	8 to 10	High-priority traveling habitat	406,624	4.6
8 to 10	1 to 7	High priority encamped habitat	406,624	4.6
8 to 10	8 to 10	High priority encamped and traveling habitat	2,240,132	25.4
1 to 7	1 to 7	Lower priority habitat	5,769,140	65.4
Total			8,822,520	

Raster surfaces depicting encamped and traveling behavior of greater sage-grouse (*Centrocercus urophasianus*) in south-central Wyoming, USA, during 2009 to 2012 were developed with the predicted probability of occurrence classified into ten relative probability bins (1 = lowest, 10 = highest) that included 10% of the landscape area. The encamped and traveling surfaces were combined to provide an output raster in which a unique output value (designation) was assigned to each of four combinations of input values.

Dzialak *et al.*

Dzialak *et al.* *Ecological Processes* 2015 4:5, doi: 10.1186/s13717-015-0032-2

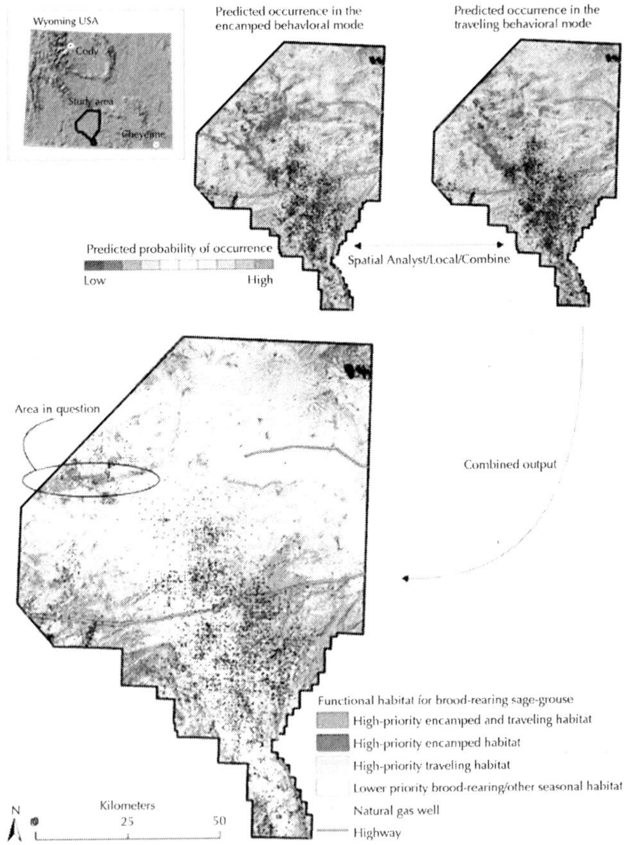

Figure 3: Study area and spatial depiction of occurrence by greater sage-grouse in Wyoming. Spatial depiction of occurrence of greater sage-grouse (*Centrocercus urophasianus*) in south-central Wyoming, USA, from 2009 to 2012 estimated from discrete-choice models of encamped and traveling behavioral modes. The combined output depicts landscapes that serve concurrently in high-priority encamped and traveling modes (25.4%; see Table 7), as well as landscapes that are used for either encampment (4.6%) or traveling (4.6%), but not both (65.4%). The 'area in question' is characterized by sand dunes and shrub brush species and is discussed in the 'Results' section.

We note an area in question associated with these raster surfaces (Figure 3). Based on first-hand experience of the site, habitat in this area is best described as vegetated sand dunes and is comprised of dunes interspersed with vegetated hummocks that include species such as shadscale, spiny hopsage (*Grayia spinosa*), greasewood, Gardner's saltbush (*Atriplex gardneri*), and grasses. Although sage-grouse under observation have occurred in this area, its inclusion here as high-priority brood-rearing habitat may be erroneous, reflecting relatively high values for percent shrub in the covariate raster set (Table 1; Homer et al. [2012]).

Lastly, upon undertaking this investigation, we intended to analyze fatality among brooded females relative to landscape features and behavioral mode (*sensu* Dzialak et al. [2011b]); however, only six fatalities occurred during the investigation. Notably, the GPS location immediately preceding fatality (and in some cases, many locations preceding fatality) in all six instances was assigned to the encamped behavioral mode.

DISCUSSION

From these data, it appears that resource selection by sage-grouse is structured, at least partially, during the brood-rearing season, by different movement behaviors. Only one model (i.e., vegetation and water resources) was identified as the most plausible model across all scenarios. We also found that the non-behaviorally adjusted model validated well. Despite these findings, there were differences in selection for some resources in the two different behavioral modes, which highlights the importance of considering behavioral mode when investigating sage-grouse brood-rearing habitat selection. Generally, the non-behaviorally adjusted model was similar to the encamped model but had a lower weight of evidence as being the best model in the candidate set. The similarity between the encamped and non-behaviorally adjusted model could have been an artifact of the larger sample size for encamped locations; however, we accounted for this fact by weighting the data prior to analysis. We found that resource selection during traveling exhibits a much different pattern of selection because these landscape features serve a different purpose compared to the general selection patterns of sage-grouse that occur for a wide-range of behaviors (e.g., resting, encamped, foraging, etc.). Therefore, prioritization of habitat for brood-rearing sage-grouse not only needs to consider general selection patterns and demographics but also should incorporate behavior-specific patterns of selection, which may disproportionately influence population persistence or connectivity because these landscape features may not be as predominant as other features that are typically viewed as important to general resource selection of the species.

Brood-rearing in sage-grouse has been investigated in terms of days post-hatching, with the brood-rearing phase and accompanying analyses partitioned into early- and late-summer periods (Connelly et al. [2000]; Thompson et al. [2006]), and the landscape context within which these resources occur establish a reasonable basis for investigating temporal patterns of selection (Mysterud and Ims [1998]; Sveum et al. [1998]; Thompson et al. [2006]; Dzialak et al. [2011b]) yet do not describe how resources are used during different behaviors. Another way to look at brood-rearing involves the behavioral mode underpinning observed movement and occurrence patterns. At a fundamental level, within- versus between-patch movements are dominant behavioral modes among female sage-grouse during brood-rearing. The fine-scale temporal data provided by GPS units made this apparent (Figure 1). Behavioral mode, as a latent process influencing occurrence patterns, reflects spatial and temporal attributes of variation in local conditions driving resource availability independently of days post-hatching. The estimated relative displacement index was one approach to incorporating information on behavior to strengthen inference on resource needs (Figures 1 and 2). Other approaches to identify behavioral modes or to assign data to defined modes have involved random walk models (Wu et al. [2000]), cluster analysis (Van Moorter et al. [2010]), state-space models (Jonsen et al. [2005]), fractals (Etzenhouser et al. [1998]; Webb et al. [2009]), and generalized additive models (Dzialak et al. [2011a]). Results of the resource selection models, including their validation metrics, indicate that the relative displacement index assigned data to the respective behavioral modes in a biologically meaningful way.

For both encamped and traveling modes and the non-behaviorally adjusted model, the importance of vegetation attributes and proximity to mesic areas was the best model among the candidate models. This suggests that all three models shared key attributes, perhaps most notably the importance of mesic conditions. Such overlap also can be seen in the extensive area that was used concurrently in high-priority encamped and traveling modes (Table 7, Figure 3). Nonetheless, the sign and magnitude of effect of key parameters differed between modes, and compared to the non-behaviorally adjusted model, in ways that may broaden our view of what important brood-rearing habitat looks like. The selection patterns during the traveling mode changed for 6 of 13 (46.2%) variables included in the model compared to the encamped and non-behaviorally adjusted model, which both had similar patterns of selection. The analysis also identified some areas that were suitable for either encampment or traveling modes, but not both (Table 7, Figure 3), suggesting some degree of divergence between the habitats. Assuming the encamped mode involved within-patch foraging, habitat used for forage resources was characterized by features that have become familiar to managers including areas with taller shrubs, proximity to mesic areas, and avoidance of bare ground (Figure 4a; Wallestad [1971]; Dunn and Braun [1986]; Drut et al. [1994]; Connelly et al. [2000]; Aldridge and Boyce [2007]; Dzialak et al. [2011b]).

Figure 4: Representative habitat for greater sage-grouse during encamped and traveling phases in Wyoming. Encamped habitat for greater sage-grouse (*Centrocercus urophasianus*) in south-central Wyoming, USA, is shown in panel (a), and traveling habitat is shown in panel (b). Landscape features used for encampment behaviors included shrub coverage with a prominent sagebrush

component, tall shrubs, limited bare ground, and proximity to mesic areas. Traveling habitat included limited shrub coverage, a higher proportion of herbaceous litter, and slightly diminished proximity to mesic areas. Note the shallow mesic area (albeit dry), established by snowmelt, situated within an extensive area of limited shrub coverage in panel (b). Photos by C.V. Olson.

Characterizing attributes of the landscape that were important during between-patch movement was a novel addition to our understanding of brood-rearing sage-grouse. While traveling, sage-grouse selected for proximity to mesic areas, but the strength of selection was diminished relative to encamped behavior. Notably, there was no apparent selection for sagebrush and shrubs among traveling sage-grouse (Table 5). From a management perspective, habitat used during between-patch movement can be described as mesic stopover sites interspersed throughout areas characterized by a high proportion of grasses or herbaceous litter (Figure 4b). Compared to encampment areas, which likely provided available forage and cover, traveling sage-grouse tended to be influenced more by the amount of litter and herbaceous vegetation. The use of open areas (i.e., less shrub coverage) in close proximity to mesic areas also could be an advantage to reduce predation whereby sage-grouse can visually detect potential predators (Webb et al. [2012]). The link between behavioral modes and the reason why a given area was selected during both encamped and traveling modes is the importance of mesic areas to brooded females, chicks, and the survival of both. Although we could not link mortality with selection of resources because of limited sample size, it is noteworthy that all fatalities were classified to an encamped location. Although speculative, this finding also supports the selection of resources during traveling that may minimize predation by using open areas (i.e., for visual detection of predators) in close proximity to mesic areas that could provide hiding cover.

Models that included anthropogenic features, terrain, and topographic features performed poorly, capturing far less information in the data than the top model. Aldridge and Boyce ([2007]) found that broods avoided areas with a high density of human development, including oil and natural gas wells. In a nearby Wyoming study area, we found consistent avoidance of rough terrain at the patch scale (90 m^2) among brooded females (Dzialak et al. [2011b]). Companion studies in this and nearby study areas found that the specific response of sage-grouse to anthropogenic and other landscape features was mediated by behavioral mode. Certain behavior such as nest site (and possibly winter range) selection showed relative generality across regions, whereas other behavior such as nesting off-bout occurrence showed a higher degree of region-specificity (Dzialak et al.[2013a]; [2013b]). A meaningful relationship between brooded sage-grouse and anthropogenic or terrain features likely exists in this area, with different analytical approaches among studies (i.e., information theoretic versus variable reduction) contributing to differing inference. But it is also plausible that resource acquisition during brood-rearing is highly adaptable to spatial and temporal patterns in floristic and predatory composition of the local area such that inference on the relationship between brood occurrence and particular landscape features should be investigated regionally.

CONCLUSIONS

Most overlap in the selection for resources occurred between encamped and non-behaviorally adjusted models, although selection patterns for some features, primarily mesic areas, were similar across all behaviors and models. These results draw attention to the key brood-rearing process of between-patch movement and to the range of landscape features that play a role in this process. The combined raster surface (Figure 3) makes the analytically derived characteristics of habitat spatial. This surface validated well and offers a reliable habitat management tool that is readily amenable to application in efforts to focus sustainable landscape planning. As a GIS product, the raster surface can be included in infrastructure siting, habitat avoidance, and reclamation processes by GIS users with the managing agencies and industry. In lower-elevation Intermountain landscapes, mesic conditions that establish important brood-rearing patches often arise through deposition of water for agriculture or are produced as a byproduct of energy development. While the use of produced water may create brood-rearing habitat, there is evidence that such habitat, when in proximity to infrastructure, results in an ecological trap by establishing resource subsidies that modify predator abundance or effectiveness in ways that are detrimental to brooded females (see Dzialak et al. [2011b]; Webb et al. [2012]). Potential encampment sites may only be used when connected by suitable travel corridors composed of unique resources. If suitable travel corridors were not available, then potential encampment sites may no longer function in the persistence of populations because animals may not be able to reach encamped sites or may face greater risk when traveling to these sites. Therefore, a relevant aim of future research would be to further characterize habitat in terms of connectivity (Harju et al. [2013b]), stopover sites (Sawyer et al. [2009]), and to more specifically quantify risk associated with behavioral modes or activities (e.g., during encampment or traveling).

AUTHORS' CONTRIBUTIONS

MRD designed the study, provided statistical assistance, analyzed portions of the data set, and helped to draft the manuscript. CVO designed the study, managed data, and helped to draft the manuscript. SLW analyzed data and helped to draft the manuscript. SMH provided statistical analysis and helped to draft the manuscript. JBW designed the study and reviewed manuscript drafts. All authors have read and approved the manuscript.

ACKNOWLEDGEMENTS

Funding was provided by BP America. We thank C. Hedley, A. Gregory, and several anonymous reviewers for providing insightful comments on early versions of this manuscript.

REFERENCES

1. Aldridge CL, Boyce MS: Linking occurrence and fitness to persistence: habitat based approach for endangered greater sage-grouse. *Ecol Appl* 2007, 17:508-526.
2. Aldridge CL, Brigham RM: Sage-grouse nesting and brood habitat use in southern Canada. *J Wildl Manage* 2002, 66:433-444.
3. Arnold TW: Uninformative parameters and model selection using Akaike's information criterion.*J Wildl Manage* 2010, 74:1175-1178.

4. Baasch DM, Tyre AJ, Millspaugh JJ, Hygnstrom SE, Vercauteren KC: An evaluation of three statistical methods used to model resource selection. *Ecol Model* 2010, 221:565-574.

5. Burnham KP, Anderson DR: *Model selection and multimodel inference: a practical information theoretic approach.* 2nd edition. Springer, New York; 2004.

6. Compton BW, Rhymer JM, McCollough M: Habitat selection by wood turtles (*Clemmys insculpta*): an application of paired logistic regression. *Ecol* 2002, 83:833-843.

7. Connelly JW, Braun CE: Long-term changes in sage grouse *Centrocercus urophasianus*populations in western North America. *Wildl Biol* 1997, 3:229-234.

8. Connelly JW, Schroeder MA, Sands AR, Braun CE: Guidelines to manage sage grouse populations and their habitats. *Wildl Soc Bull* 2000, 28:967-985.

9. Connelly JW, Knick ST, Schroeder MA, Stiver SJ: *Conservation assessment of greater sage-grouse and sagebrush habitats.* Western Association of Fish and Wildlife Agencies, Cheyenne, Wyoming; 2004.

10. Cooper AB, Millspaugh JJ: The application of discrete choice models to wildlife resource selection studies. *Ecol* 1999, 80:566-575.

11. COSEWIC: *COSEWIC assessment and update status report on the greater sage-grouse Centrocercus urophasianus, Phaios subspecies and Urophasianus subspecies, Centrocercus urophasianus urophasianus, in Canada*. Committee on the Status of Endangered Wildlife in Canada, Ottawa, Canada; 2008.

12. Crunden CW: Age and sex of sage-grouse from wings. *J Wildl Manage* 1963, 27:846-849.

13. Cushman SA, Lewis JS: Movement behavior explains genetic differentiation in American black bears. *Landsc Ecol* 2010, 25:1613-1625.

14. Doherty KE, Naugle DE, Walker BL, Graham JM: Greater sage-grouse winter habitat selection and energy development. *J Wildl Manage* 2008, 72:187-195.

15. Doherty KE, Naugle DE, Walker BL: Greater sage-grouse nesting habitat: the importance of managing at multiple scales. *J Wildl*

Manage 2010, 74:1544-1553.

16. Drut MS, Pyle WH, Crawford JA: Diets and food selection of sage grouse chicks in Oregon. *Range Manage* 1994, 47:90-93.
17. Dunn PO, Braun CE: Summer habitat use by adult female and juvenile sage-grouse. *J Wildl Manage* 1986, 50:228-235.
18. Dzialak MR, Harju SM, Osborn RG, Wondzell J, Hayden-Wing LD, Winstead JB, *et al.*:Prioritizing conservation of ungulate calving resources in multiple use landscapes. *PLoS One* 2011, 6(1):e14597.
19. Dzialak MR, Olson CV, Harju SM, Webb SL, Mudd JP, Winstead JB, *et al.*: Identifying and prioritizing greater-sage grouse nesting and brood-rearing habitat for conservation in human-modified landscapes. *PLoS One* 2011, 6(10):e26273.
20. Dzialak MR, Olson CV, Harju SM, Webb SL, Winstead JB: Temporal and hierarchical spatial components of animal occurrence: conserving seasonal habitat for greater sage-grouse. *Ecosphere* 2012, 3(3):30.
21. Dzialak MR, Olson CV, Harju SM, Winstead JB: Spatial generality of predicted occurrence models of nesting habitat for the greater sage-grouse. *Ecosphere* 2013, 4(3):41
22. Dzialak MR, Webb SL, Harju SM, Olson CV, Winstead JB, Hayden-Wing LD: Greater sage-grouse and severe winter conditions: identifying habitat for conservation. *Rangeland Ecol Manage* 2013, 66:10-18.
23. Etzenhouser MJ, Owens MK, Spalinger DE, Murden SB: Foraging behavior of browsing ruminants in a heterogeneous landscape. *Landsc Ecol* 1998, 13:55-64.
24. Fedy BC, Aldridge CL: The importance of within-year repeated counts and the influence of scale on long-term monitoring of sage-grouse. *J Wildl Manage* 2011, 75:1022-1033.
25. Forester JD, Im HK, Rathouz PJ: Accounting for animal movement in estimation of resource selection functions: sampling and data analysis. *Ecol* 2009, 90:3554-3565.
26. Guisan A, Thuiller W: Predicting species distribution: offering more than simple habitat models. *Ecol Lett* 2005, 8:993-1009.
27. Hagen CA, Connelly JW, Schroeder MA: A meta-analysis of greater sage-grouse*Centrocercus urophasianus* nesting and

brood-rearing habitats. *Wildl Biol* 2007, 13:42-50.

28. Harju SM, Dzialak MR, Taylor RC, Hayden-Wing LD, Winstead JB: Thresholds and time lags in the effects of energy development on greater sage-grouse populations. *J Wildl Manage* 2010, 74:437-448.
29. Harju SM, Olson CV, Foy-Martin L, Webb SL, Dzialak MR, Winstead JB, *et al.*: Occurrence and success of greater sage-grouse broods in relation to insect-vegetation community gradients.*Hum-Wildl Interact* 2013, 7:214-229.
30. Harju SM, Olson CV, Dzialak MR, Mudd JP, Winstead JB: A flexible approach for assessing functional landscape connectivity, with application to greater sage-grouse (*Centrocercus urophasianus*). *PLoS One* 2013, 8(12):e82271.
31. Holloran MJ, Kaiser RC, Hubert WA: Yearling greater sage-grouse response to energy development in Wyoming. *J Wildl Manage* 2010, 74:65-72.
32. Homer CG, Aldridge CL, Meyer DK, Schell SJ: Multi-scale remote sensing sagebrush characterization with regression trees over Wyoming, USA: laying a foundation for monitoring. *Int J Appl Earth Obs Geoinf* 2012, 14:233-244.
33. Johnson CJ, Seip DR, Boyce MS: A quantitative approach to conservation planning: using resource selection functions to map the distribution of mountain caribou at multiple spatial scales. *J Appl Ecol* 2004, 41:238-251.
34. Jonsen ID, Flemming JM, Myers RA: Robust state-space modeling of animal movement data. *Ecol* 2005, 86:2874-2880.
35. Kindlmann P, Burel F: Connectivity measures: a review. *Landsc Ecol* 2008, 23:879-890.
36. Manly BFJ, McDonald LL, Thomas DL, McDonald TL, Erickson WP: *Resource selection by animals: statistical design and analysis for field studies*. 2nd edition. Kluwer Academic Publishers, Norwell, Massachusetts; 2002.
37. Margules CR, Pressey RL: Systematic conservation planning. *Nature* 2000, 405:243-253.
38. McCune B, Keon D: Equations for potential annual direct incident radiation and heat load. *J Veg Sci* 2002, 13:603-606

39. McDonald TL, Manly BFJ, Nielson RM, Diller LV: Discrete-choice modeling in wildlife studies exemplified by northern spotted owl nighttime habitat selection. *J Wildl Manage* 2006, 70:375-383.

40. Moilanen A, Franco AMA, Early RI, Fox R, Wintle B, and Thomas CD: Prioritizing multiple-use landscapes for conservation: methods for large multi-species planning problems. *P Roy Soc Lond B Bio* 2005, 272:1885-1891.

41. Morales JM, Moorcroft PR, Matthiopoulos J, Frair JL, Kie JG, Powell RA, *et al.*: Building the bridge between animal movement and population dynamics. *Philos T Roy Soc B* 2010, 365:2289-2301

42. Mysterud A, Ims RA: Functional responses in habitat use: availability influences relative use in trade-off situations. *Ecol* 1998, 79:1435-1441.

43. Patterson RL: *The sage grouse in Wyoming*. Sage Books, Denver, Colorado; 1952.

44. Patterson TA, Thomas L, Wilcox C, Ovaskaien O, Matthiopoulos J: State-space models of individual animal movement. *Trends Ecol Evol* 2008, 23:87-94.

45. Pendergast JF, Gange SJ, Newton MA, Lindstrom MJ, Palta M, Fisher MR: A survey of methods for analyzing clustered binary response data. *Int Stat Rev* 1996, 64:89-118.

46. Sawyer H, Kauffman MJ, Nielsen RM, Horne JS: Identifying and prioritizing ungulate migration routes for landscape-level conservation. *Ecol Appl* 2009, 19:2016-2025.

47. Sveum CM, Crawford JA, Edge WD: Use and selection of brood-rearing habitat by sage-grouse in south central Washington. *Great Basin Nat* 1998, 58:344-351.

48. Taylor PD, Fahrig L, Henein K, Merriam G: Connectivity is a vital element of landscape structure. *Oikos* 1993, 68:571-572.

49. Thompson KM, Holloran MJ, Slater SJ, Kuipers JL, Anderson SH: Early brood-rearing habitat use and productivity of greater sage-grouse in Wyoming. *West N Am Naturalist* 2006, 66:332-342.

50. Tischendorf L, Fahrig L: On the usage and measurement of landscape connectivity. *Oikos* 2000, 90:7-19.

51. Turchin P: *Quantitative analysis of movement: measuring and*

modeling population redistribution in plants and animals. Sinauer Associates, Sunderland, Massachusetts; 1998.

52. United States Fish and Wildlife Service: Endangered and threatened wildlife and plants: 12-month finding for petitions to list the greater sage-grouse (*Centrocercus urophasianus*) as threatened or endangered *Fed Regist* 2010, 75:13910-14014.
53. Van Moorter B, Visscher DR, Jerde CL, Frair JL, Merrill EH: Identifying movement states from location data using cluster analysis. *J Wildl Manage* 2010, 74:588-594.
54. Visual Learning Systems: *Feature analyst 4.2 for ArcGIS reference manual*. Visual Learning Systems, Missoula, Montana; 2008.
55. Wakkinen WL, Reese KP, Connelly JW, Fischer RA: An improved spotlighting technique for capturing sage-grouse. *Wildl Soc Bull* 1992, 20:425-426.
56. Walker BL, Naugle DE, Doherty KE: Greater-sage-grouse population response to energy development and habitat loss. *J Wildl Manage* 2007, 71:2644-2654.
57. Wallestad RO: Summer movements and habitat use by sage grouse broods in central Montana. *J Wildl Manage* 1971, 35:129-135.
58. Webb SL, Riffell SK, Gee KL, Demarais S: Using fractal analyses to characterize movement paths in white-tailed deer and response to spatial scale. *J Mammal* 2009, 90:1210-1217
59. Webb SL, Olson CV, Dzialak MR, Harju SM, Winstead JB, Lockman D: Landscape features and weather influence nest survival of a ground-nesting bird of conservation concern, the greater sage-grouse, in human-altered environments. *Ecol Process* 2012, 1:4. BioMed
60. Western Regional Climate Center (2011) Period of record monthly climate summary, Rawlins FAA Airport, Wyoming. Western Regional Climate Center, Reno, Nevada. http://www.wrcc.dri.edu/cgi-bin/cliMAIN.pl?wy7533. Accessed 20 November 2014
61. Wu H, Li B, Springer TA, Neill WH: Modelling animal movement as a persistent random walk in two dimensions: expected magnitude of net displacement. *Ecol Model* 2000, 132:115-124.

3

Mountain Pastoralism in Transition: Consequences of Legalizing cordyceps Collection on Yak Farming Practices in Bhutan

Kesang Wangchuk and Jigme Wangdi

Renewable Natural Resources Research Center, Ministry of Agriculture and Forests, Bumthang 32001, Bhutan

ABSTRACT

Yak farming is the main livelihood source for the high altitude communities in the eastern Himalaya. With increasing access to modern facilities, market opportunities and changes in the legal framework, pastoral systems in the Himalaya are undergoing an unprecedented change. Questionnaire-based qualitative surveys were conducted in five villages of northern Bhutan, to understand how the recent changes in the legal framework for *Cordyceps* (known as caterpillar fungus) collection have caused specific changes in yak farming practices.

Survey results revealed that women were increasingly involved in yak husbandry and household work, after the legalization of *Cordyceps* collection in 2004. After legalization, the *Cordyceps* business overtook yak farming as the main income-earning activity. Post-legalization saw a decline in the overall grassland condition and most herders migrated a month earlier to the summer grazing land. Legalization also led to increase in the number of households buying commercial feeds for yaks. Yak mortality increased and fodder scarcity became more acute, which is a major constraint to yak farming. Despite the good income from the *Cordyceps* business, yak farming was the preferred earning activity over *Cordyceps* due to herders' confidence in yak farming as a reliable source of livelihood. Of several measures proposed by yak herders to improve yak farming, increasing grassland productivity and providing subsidies for feed purchases were the most important measures. The study concluded that yak farming practices have undergone a few positive but more undesirable changes after the legalization of *Cordyceps* collection in 2004. The results suggest multi-disciplinary approaches to address adequately the emerging issues of yak farming e.g. introducing schemes to make yak farming attractive to the mountain youth. The paper suggests interventions to strengthen yak farming and help herders make informed choices in the high altitude rangelands of Bhutan. Essentially, yak farming is at a crossroads where a firm decision is needed to either encourage and strengthen the farming practices or witness the gradual extinction of the age-old tradition.

BACKGROUND

Mountains occupy one fifth of the earth's surface and are home to at least 1.2 billion people (Körner and Ohsawa [2005]). Inaccessibility and marginality make mountains one of the toughest environments for agriculture in terms of diversification of farm activities, resulting in a higher prevalence of vulnerability and warranting particular attention for sustainability of the overall mountain systems (Körner and Ohsawa [2005]). In spite of the complex challenges, pastoralists adopt suitable strategies to sustain their livelihoods, with their ability to transform the extensive marginal rangelands into economically productive areas (Mishra et al. [2010]). Pastoralists have sustained their economies through efficient utilization of grassland resources by high altitude livestock species, ranging from camelids in the Andes to yak *(Bos grunniens)* in the Himalayan mountains. However, pastoralism is evolving more rapidly due to changes in the socio-economic, institutional and policy environments (Shreshtha [1994]). Several authors (Roder et al. [2002]; Tulachan and Neupane [1999]; Wangchuk et al. [2014]) note that socio-economic development is the main driver of change in the Himalayan farming systems. To deepen our understanding of transitions in mountain farming in modern times, we investigated the case of yak farming in the Bhutan Himalaya, where farming practices have undergone an unprecedented change after the legalization of *Cordyceps* (*Ophiocordyceps sinensis*) collection for medicinal purposes.

Yak farming is predominant and a major livelihood source for the high altitude communities in Bhutan. However, recently yak farming has become marginalized and overtaken by the market demand for *Cordyceps*or the caterpillar fungus. *Cordyceps,* a restricted species until 2003, was legalized for collection in 2004, and has become a major source of income for the high altitude communities (Wangmo and Wangchuk [2008]). *Cordyceps* are found in the remote northern parts of Bhutan, at an altitude of 3,400-4,100 m above sea level (Wangmo and Wangchuk [2008]). The sale of *Cordyceps* fetches herding households a good annual income of up to US$ 24,500 (Wangchuk et al. [2012]). Internationally, the value of *Cordyceps* was reported to have increased by over 1,000% between 1997 and 2012 (Winkler [2013]). However, economic development has taken place in the mountains at a price. Environmental pollution through littering and grassland degradation are rampant (Wangchuk et al. [2012]; Wangmo and Wangchuk [2008]), and may consequently degrade the ecosystem services. Economic development in the mountains is also an indirect factor that weakens traditional cultures (Körner and Ohsawa [2005]).

To date, empirical studies on *Cordyceps*have focused mainly on the medicinal properties (Boesi and Cardi [2009]; Miller [2009]; Panda and Swain [2011]), sustainable management (Cannon et al. [2009]; Tenzin [2009]; Wangmo and Wangchuk [2008]) and impacts on livelihood and environment (Wangchuk et al. [2012]). A study investigating the effects of legalization of*Cordyceps* collection on yak farming practices was therefore needed. Wangmo and Wangchuk ([2008]) provide anecdotal evidence suggesting that *Cordyceps* collection may compete with yak farming, change yak farming practices and transform the lifestyle of herding communities. Despite these concerns, the changes remain unacknowledged and herding communities do not raise the issue for fear of inviting government policies banning the *Cordyceps* collection. Hence, there is little understanding of the long-term consequences of indiscriminate and unsustainable collection of*Cordyceps* to an age-old culture and farming environment of these rangeland communities.

Policies to promote livestock farming require sufficient knowledge of the existing production systems and changing practices, without which there could be highly diverse consequences on farm management and future evolution. Without this knowledge, it is highly likely that policy decisions will become difficult and impending conflicts more complex to resolve. Therefore, the objectives of this study were to understand how changes in the legal framework for *Cordyceps*collection have caused changes in yak farming practices in the high altitude rangelands of Bhutan, and to suggest future interventions for harmonious development of yak herding communities in view of *Cordyceps* collection activities. The paper unravels the consequences arising from*Cordyceps* collection activities, contributes to our understandings of the farming scenario and provides insights to policy decisions to benefit the yak herding communities in the longer run.

STUDY AREAS

The main *Cordyceps* collection areas are found in north eastern and western Bhutan. The villages of Naro, Lingzhi, SoiYaksa, Sephu and Khangdang were selected for the study (Figure 1).

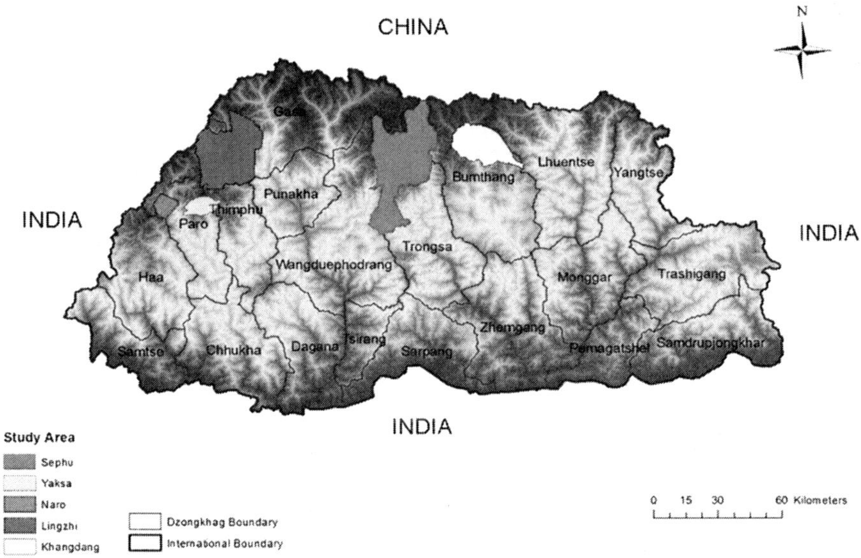

Figure 1: Study sites in northern Bhutan where *Cordyceps* collection and yak farming are predominant.

The villages are far flung and about two days' walk from the nearest motor-able roads.

The topography of study villages is rugged and the mean maximum temperature of 20°C is recorded in summer and the mean minimum of −10°C in winter. Summer is moderately warm and wet followed by freezing cold and dry winters. The annual rainfall shows a unimodal distribution pattern, with the maximum precipitation from June to September.

Every year, huge numbers of people spend about a month from mid-May to mid-June, collecting*Cordyceps* in the high altitude environment. Collection is banned in other months of the year. To prevent overexploitation of resource, the collection rules were revised within short intervals. Until 2007, only one person from a household was allowed to collect *Cordyceps*, however, in 2008, the revised rule allowed all members from a household to collect (Cannon et al. [2009]; Gould [2007]; Wangchuk [2008]). From 2009 to date, a new rule was enforced, allowing only three persons from each household to collect. Many of the *Cordyceps* growing sites in Bhutan are near the border areas of Tibet. This makes easier for the people from the other side of the border to trespass and illegally collect *Cordyceps* in Bhutan. The practice continues unabated, even with the strict monitoring by the forestry and security personnel.

METHODS

The study was conducted in the summer of 2014 in selected high altitude areas of Thimphu, Paro, Wangdue and Bumthang districts (Figure 1), where yak farming and *Cordyceps* collection are major income generating activities. For selecting villages and respondents, we employed the snowball sampling technique (Biernacki and Waldorf [1981]) at two stages. In the first stage, researchers, development workers and local government bodies were consulted who provided a list of villages actively involved in *Cordyceps* business and yak farming. In the second stage, village leaders were consulted who recommended households and elderly herders as potential respondents of survey interviews.

Field Interviews

Using the snowball sampling technique (Biernacki and Waldorf [1981]), a total of 100 older herders were identified, of who 60% were male. Both male (50–70 years of age) and female (50–65 years) herders from five villages were recruited as key informants since they were actively involved in yak farming over the last three decades and were able to provide farming details of the past, present and changes likely in the future. The selection criteria included: the herder engaged in yak farming for the past 30 years, have witnessed change in farming practices overtime, be a permanent resident of the village, and be willing to cooperate with the interviewer.

The survey was administered through questionnaires with open- and closed-ended questions. The survey questionnaire was divided into two parts. The first part consisted of questions on the overall situation 10 years before the legalization of *Cordyceps* collection. The same questions were repeated in the second part of the questionnaire in order to track changes in farming practices10 years after the legalization. Information, missed by the questionnaire was captured through informal interviews and discussions during tea breaks and mealtimes after survey interviews. Topics and questions during informal discussions and interviews included hardships of yak farming, rural out-migration, incomes from yak farming and the *Cordyceps* business, and migratory and feeding practices for yaks. Informal discussions and interviews lasted from 30 minutes to one hour.

The first subpart of the questionnaire solicited information on the role of gender and socioeconomics of the herding community. The key informants were asked about the role of gender in yak husbandry. To better understand the changes in the population of the younger generation, the key informants were asked to provide opinions on population size before and after the legalization of *Cordyceps* collection. Questions were also asked to quantify changes in the number of households and herds. The key informants were requested to rank the income sources before and after the legalization of *Cordyceps* collection.

The second part of the questionnaire delved into grazing site condition and yak feeding practices. To simplify information, we used a three-point Likert scale (Likert [1932]) against the commonly used five-point scale, since reliability and validity of information was found to be independent of the scale points employed by Likert-type items (Jacoby and Miichael [1971]). The scale was used to gather information on whether the informants agreed, disagreed or did not know about the changes in grazing site condition, extent of forage oat cultivation and feeding of commercial animal feeds to yak, before and after the legalization of *Cordyceps* collection. We requested information on migratory practices, mainly on the time of migration to summer and winter grazing lands. This part of the questionnaire captured information on the yak mortality rate and its main causes, before and after the legalization of *Cordyceps* collection. The third part of the questionnaire included questions aimed at identifying the major constraints to yak farming. The constraints before and after the legalization of *Cordyceps* collection were compared.

The final part of the questionnaire sought general opinions on whether the herdsmen preferred yak farming or *Cordyceps* business. Accordingly, reasons were gathered to explain the preference for particular income-generating activities. The key informants were requested to opine on whether yak farming would decline in the future, followed by questions requesting to substantiate their opinions. Finally, the key informants were asked to put forward their important requests to the government, in order to improve and encourage yak farming in the high altitude rangeland.

Data Analysis

Depending on the type of data, both parametric and nonparametric tests were performed. The paired-samples t tests were performed on quantitative data. Opinions were displayed by percentages and the chi-square tests were conducted to evaluate differences in opinions before and after the legalization of *Cordyceps* collection. Ranks for income sources were analyzed with Wilcoxon signed-rank test. The entire dataset was analyzed with SPSS 22 (Landau and Everitt[2004]). Comparisons were not made between villages since our interest was to understand the overall changes in yak farming practices before and after the legalization of *Cordyceps* collection.

RESULTS AND DISCUSSION

Gender, Livestock and Socio-Economics

The percentage of family heads, yak herders by gender and the population size of younger generation, before and after the legalization of *Cordyceps* collection, are presented in Table 1. The results showed a significant decline in the proportion of men heading the family and herding yaks to and from the grazing fields, whereas the proportion of women as head of family and involvement in yak herding increased significantly after the legalization of *Cordyceps* collection. The proportion of family heads and yak herders differed by gender, which is consistent with the finding of Joshi et al. ([2013]) that women are increasingly involved in rangeland resource management in the Hindu Kush Himalayan region. Following legalization of *Cordyceps* collection, women are being increasingly involved in animal husbandry and household work, suggesting the growing leadership of women in domestic affairs. This is a positive development, indicating considerable improvement in women's role at the family level, suggesting that women might have more control over yak farming in the future. It would also mean that such positive developments may overcome gender-based constraints hindering women as farmers and managers of natural resources, particularly from the perspective of the future of yak farming. However, the bigger role of women would also mean increase in responsibilities and labour burden of women. Although, this study was unable to generate evidence on how men devote most of their time to off-farm activities such as *Cordyceps* collection, tourism and market related activities, Joshi et al. ([2013]) reported that men's roles in the Himalayan region are limited to their involvement in the wage labour and marketing of agricultural products and they spend less time on pastoral activities, as compared with women. Therefore, the results suggest that men shall have a greater role in off-farm income generating activities in the future. The findings of this study on the changing role of gender should not be considered as a general trend but mainly a phenomenon in yak farming and *Cordyceps*growing regions in the Bhutan Himalaya.

Table 1: Gender role and size of younger generation before (1993–2003) and after (2004–2014) the legalization of *cordyceps* collection

	Before legalizing cordyceps collection (%) (Respondents N = 100)	After legalizing cordyceps collection (%) (Respondents N = 100)	Sig.
Head of family			
Men	63	55	**
Women	37	45	**
Yak herder			
Men	34	28	*
Women	57	62	*
Children	9	10	ns
Population size of younger generation			
Small	30	62	**
Medium	25	14	**
Big	45	14	**
Number of households	837	989	ns

Number of herds	591	516	ns

*p≤0.05, **p≤0.01, ns-non-significant.

Wangchuk and Wangdi

Wangchuk and Wangdi Pastoralism 2015 5:4, doi:10.1186/s13570-015-0025-x

The population size of the younger generation was significantly smaller after the legalization of*Cordyceps* collection (Table 1), showing a decline in the population of rural youth. This is likely to result in acute shortage of farm labourers and may impede yak farming in the future. Informal discussions with both young and old villagers revealed that yak farming is becoming increasingly difficult. The younger generation are aware of the hardship of yak husbandry in remote and inaccessible areas and the low return on investment as compared to the off-farm businesses such as tourism and *Cordyceps* collection. Thus, yak farming is becoming less popular with the younger generation. Although, the causes of rural youth decline are multiple and complex, Wangchuk et al. ([2014]) attributed rural–urban migration to the rising tendency of young and literate people to migrate to urban areas seeking better economic opportunities. Relating the current trend of declining number of youth to the future population structure, our study suggests that the future herding community will most likely be comprised of elderly people. This cautions the need for the herding communities to adapt to the generational change. Accordingly, it sends out an important signal for the government agencies to step up efforts to address the issue, if yak farming is not to face extinction.

Although non-significant, we detected an increase in the number of households by 152 over the last two decades and a decline in the number of herds by 75 after *Cordyceps* collection (Table 1). The increase in household numbers and decline in herd numbers may not necessarily be due to legalization but can be attributed to two other possible reasons. Firstly, the increase in household numbers is probably attributable to the customary practice of traditional rural families in Bhutan of dividing land between the family members, with each generation inheriting consecutively smaller patches of land (Wangchuk et al. [2014]). The new houses built on inherited land results in the increase in number of households over the years. Secondly, informal discussions revealed that, due to better economic opportunities elsewhere and increasing difficulty in yak farming, there are cases of some herding families giving up yak farming and migrating to urban areas in the recent years. The herds were often sold to fellow herders, which might explain the consequent decline in the total number of herds without a decline in overall yak numbers, while individual herd sizes may have increased.

About 77 percent of respondents ranked yak farming as the main source of income for the herding families before legalizing collection of *Cordyceps* (Figure 2). However, after 2004, with herding families legally authorized to collect *Cordyceps*, about 70 percent of respondents mentioned the collection and sale of *Cordyceps* as the main source of income, followed by yak farming. Herdsmen often claim that the income obtained from a month- long collection of *Cordyceps* is by many folds greater than the income earned from year-long yak farming. This is probably the main reason why the numbers of *Cordyceps* collectors are on the rise in the recent years (Wangchuk et al. [2012]).

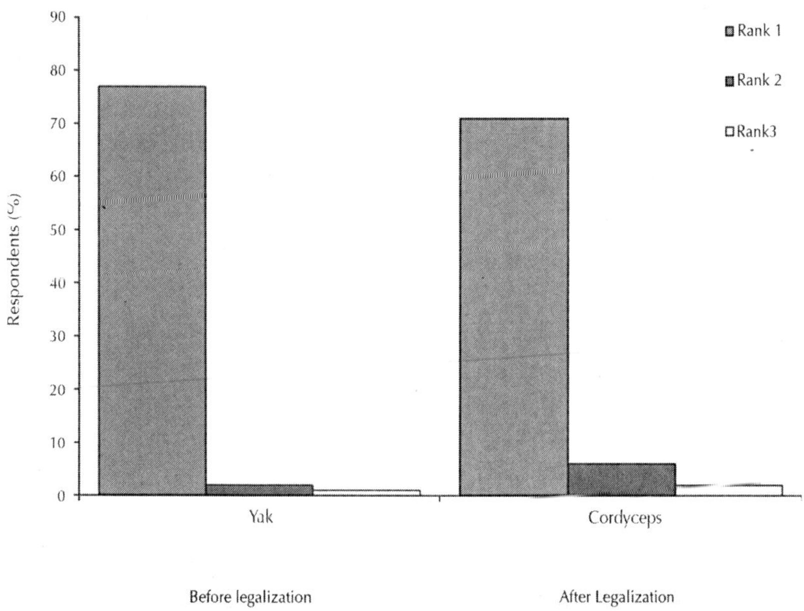

Figure 2: Main income sources before (1993–2003) and after (2004–2014) the legalization of*Cordyceps*collection.

Grassland Condition

Grazing land was said to have degraded after the legalization of *Cordyceps* collection, according to over 70 percent of the respondents (Figure 3). The results highlight the perception of increased degradation of the grasslands arising· from continued intensive collection of *Cordyceps* and overgrazing by yaks and horses. Amongst the causes of grazing land degradation, *Cordyceps*collection activity was stated as the main cause by over 50 percent of respondents. According to herdsmen during the informal discussions, the number of illegal collectors from across the Bhutan-Tibet border is unaccounted for, and was said to outnumber the Bhutanese collectors, which contributes to grazing land degradation. Wangmo and Wangchuk ([2008]) reported that *Cordyceps*collection involved digging up soil, which is often left unattended, exposing the bare soil to erosion by monsoonal rain. Similarly to findings by Wangchuk et al. ([2012]), key informants mentioned littering of non-degradable plastic food wrappers and bottles by *Cordyceps* collectors and lack of proper garbage management as one of the causes of environmental degradation.

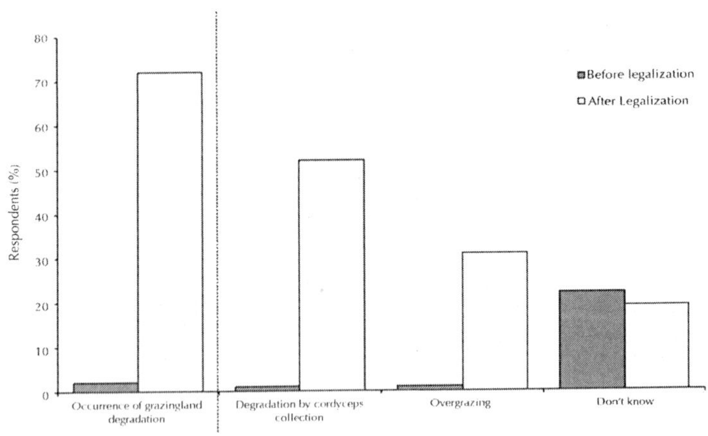

Figure 3: Occurrence and causes of poor grazing site condition before (1993-2003) and after (2004–2014) the legalization of*Cordyceps*collection.

According to a quarter of respondents, overgrazing is another cause of grassland degradation. Herders mentioned that, the *Cordyceps* collection has led to an increase in the number of horses grazing the grasslands. With the rising number of collectors over recent years, a large number of horses are used to transport basic necessities to the collection sites in order to support a month-long collection of *Cordyceps* in a difficult terrain. Furthermore, in the informal discussions, the herdsmen described the situation being aggravated by grazing pressure from horses used for tourism activities. Wangchuk et al. ([2006]) equated grazing by one horse to grazing by ten adult cattle for one day in the same area. Further, the carrying capacity of high altitude grassland is low, with one livestock unit requiring about 4.5 ha of well-managed grassland (Wangchuk et al. [2013]). Thus, overstocking on less productive grassland explains the overgrazing and consequent degradation of grassland.

Herding Practices and Constraints to Yak Farming

The yak herders migrate to summer grazing pastures in April and May, located in the high altitude region between 3,800-4,500 m above sea level; these are also the sites for *Cordyceps* collection. Although, there was no significant change in timing of migration to the winter grazing land, we detected a significant difference in migration timing to the summer grazing land, before and after the legalization of *Cordyceps* collection (Figure 4). The proportion of herders migrating to the summer grazing land in April increased by about 20 percent after the legalization. On the contrary, we detected a significant drop of about 25 percent in the proportion of herders migrating to the summer grazing land in May. The shift in time of migration suggests that most herders are spending more time in the summer grazing land, after the legalization. Since herders found that summer grazing lands are increasingly degraded by *Cordyceps* collection activities with consequent decrease in the productivity of grasslands, to ensure sufficient forage for their yaks, most herders migrate to the summer grazing land a month earlier, after legalization of *Cordyceps* collection, as opposed to the migration time followed before legalization.

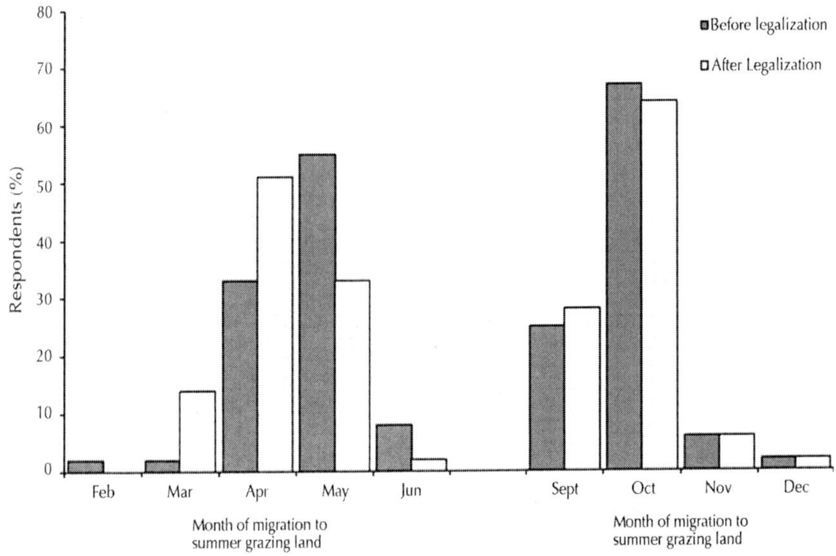

Figure 4: Month of migration to summer and winter grazing lands before (1993–2003) and after (2004–2014) the legalization of*Cordyceps* collection.

Our study showed improvements in yak feeding practices after the legalization. We detected a fivefold increase in the proportion of respondents feeding commercial feed (oil cakes and pellets) to yaks (Table 2). This practice reflects the greater purchasing power from the *Cordyceps*business, which was the main income source after the legalization. This is a positive trend, indicating that yaks are receiving better nutrition. However, about 77 percent of respondents agreed that yak mortality has increased after the legalization. Of the several reasons outlined, forage scarcity featured as the most important cause of yak mortality and a major constraint to yak farming, followed by wildlife predation and disease (*Coenurus multiceps multiceps*) (Table 2). This means that legalization and escalating pasture forage shortages have contributed to the deteriorating situation of yak farming in the Bhutan Himalaya.

Table 2: Yak mortality, feeding practice, causes of yak mortality and constraints to yak farming before (1993–2003) and after (2004–2014) the legalization ofcordyceps collection

Respondents (%)			
	Before legalizing cordyceps collection	***After legalizating cordyceps collection***	***Sig.***
1. Commercial feeds fed to yaks	20	80	***
2. Yak mortality	68	77	*
Causes of yak mortality			
1. Forage shortage	30	34	ns
2. Wildlife attack	18	20	ns
3. Gid disease	11	14	ns
4. Accidents	11	10	ns
5. Don't know	26	21	ns
Constraints to yak farming			
1. Forage shortage	68	77	*
2. Wild life attack	21	19	ns
3. Gid disease	7	7	ns
4. Don't know	5	3	ns

*$p \leq 0.05$, ***$p \leq 0.001$, ns-non-significant.

Wangchuk and Wangdi

Wangchuk and Wangdi *Pastoralism* 2015 5:4, doi:10.1186/s13570-015-0025-x

Farming Preference and Future of Yak Farming

About 63 percent of respondents preferred yak farming over the *Cordyceps* business in the long term (Table 3). This finding suggests that yak farming is an assured source of steady income for a large number of respondents, in spite of the *Cordyceps* being a lucrative business venture. When probed on the main reason for respondents' preferences, yak farming was mentioned to be sustainable and resilient at the time of food scarcity. Yaks have been reared for centuries and are strongly linked to the herders' identity and culture, which instils more confidence for yak farming. The yaks survive the harsh conditions and the herders have an intuitive ability to sustainably derive their livelihood from yak farming (Khan [1995]). On the contrary, most respondents viewed-*Cordyceps* collection as an uncertain future, due to overexploitation of the resource, fluctuation in market price and greater risks involved in collection activities in the difficult terrain under harsh weather conditions. This reiterates the prediction of Wangchuk et al. ([2006]) that the price of*Cordyceps* would fluctuate in the future. *Cordyceps*, a resource of immense commercial value, runs the risk of over-exploitation and the issue of sustainability looms large (Winkler [2013]). Reports from some parts of the *Cordyceps* growing region in China indicate a sharp decline in*Cordyceps* production due to overharvesting (Stone [2008]; Winkler [2013]). Winkler ([2013]) fears that the overly intense collection could lead to extinction of the *Cordyceps* resource. In Bhutan, without adequate monitoring of the *Cordyceps* sites, the illegal harvesting by the poachers from across the border with Tibet will threaten the sustainability of the resource.

Table 3: Main income source, preference for income activities and opinions on future yak farming

Survey question	***Herders' opinion***	***Respondents (%)***
Which income generating activity would you prefer in the long run?	1. Yak farming	63
	2. Cordyceps collection	37
Why do you prefer the particular activity?	1. Yak farming is sustainable and Cordyceps has uncertain future	63
	2. Cordyceps is sustainable	37
Will the yak farming decline in future?	1. Yak farming will decline	88
	2. Yak farming will not decline	11
	3. Don't know	01
What are the main reasons likely to cause decline in yak farming in future?	1. Yak farming is getting difficult	37
	2. Rural–urban migration	31
	3. Cordyceps collection	21
	4. Wildlife attack	06
	5. Don't know	05

Wangchuk and Wangdi

Wangchuk and Wangdi Pastoralism 2015 5:4, doi:10.1186/s13570-015-0025-x

Despite the fact that yak farming is preferred over the *Cordyceps* business, about 88 percent of respondents felt that yak farming will decline in the future (Table 3). This was linked to yak farming becoming difficult, mainly because of the increased forage shortage and yak mortality due to disease. The second important reason put forward for the likely decline in yak farming was rural out-migration. According to respondents during the informal discussions, the migration trend is on the rise and the younger generation seemed to show less interest to take up yak farming. Rural–urban migration is not only a Bhutanese phenomenon; it has been reported in most rural regions around the world due to better economic opportunities elsewhere, causing damage to rural communities in terms of skewed demographic profiles, reductions in services and loss of local culture (Stockdale [2004]). Rural out-migration is becoming common in most Asian countries, where the exit of younger family members is leaving more labour burdens on elderly people and children in rural areas, leading to rise in child farm labour (Meerza [2010]). The present scenario highlights the need to consider introducing better rural schemes and opportunities to retain rural youth and encourage yak farming. The growing interest in *Cordyceps* business shall also contribute to a decline in yak farming. This is because the income from the sale of *Cordyceps* is invested in land and infrastructure in the urban areas, where the herdsmen and their families are spending more time per year.

Future Interventions

The measures proposed by respondents to improve and encourage yak farming are presented in Table 4. Over half of the respondents felt that the improvement in yak farming can be possible through an increase in forage production, in terms of both quantity and quality, at subsidized rates. Improvements in forage production and availability may be beneficial for yak enterprises, but our study shows that the viability of yak farming will be threatened by the increasing farm labour shortage resulting from rural out-migration.

Table 4: Government interventions requested by herding communities to sustain yak farming

Sl. no.	*Measures requested by respondents to improve yak farming*	*Respondents (%)*
1	Must increase forage production and make commercial feeds available at subsidized rate	50
2	Must improve quality of breed of yak	15
3	Must introduce regulations to prevent grazing land degradation by Cordyceps collection activities	13
4	Must provide effective and timely veterinary services to control gid disease	11
5	Other minor reasons	09
6	Don't know	02

Wangchuk and Wangdi

Wangchuk and Wangdi Pastoralism 2015 5:4, doi:10.1186/s13570-015-0025-x

The need to improve the breed of yaks was the second most important request, by 15% herders who thought the yak breed is on the decline. More importance placed on the *Cordyceps* business could probably have led to less attention on improving yak breed. The decline in breed could also be explained by the repeated and traditional pure breeding practices common in the yak herding community which are reported to have harmful effects on yak performance (Wiener et al. [2003]). Inbreeding in yaks is a particular problem in Bhutan, Nepal and India (Cai and Wiener [1995]) and the lack of an efficient yak breeding program to maintain the genetic diversity in yaks is reported as a major problem in Nepal (Shaha [2000]).

The need for rules and regulations to prevent degradation of grazing land by *Cordyceps* collection activities was the third most important request to the government. The high altitude communities are fully aware of the ill effects of *Cordyceps* collection on grazing land and they expressed concern to prevent decline in grazing land productivity. Our findings confirm the apprehension of Wangmo and Wangchuk ([2008]) who found the collection techniques used by *Cordyceps* collectors destructive to grasslands.

CONCLUSIONS

Our study highlights a series of changes in yak farming practices after the legalization of*Cordyceps* collection. Legalization has resulted in a few positive but more undesirable changes in overall yak farming. Positive changes included more income from *Cordyceps* business and better nutrition for yaks. Undesirable changes were the increased role of women in household chores, deteriorated grassland condition, and increased yak mortality due to acute pasture shortage. Legislation has partly contributed to rural–urban migration. The escalating competition between yak farming and *Cordyceps* business may lead to a decline in yak farming in the long run. This may accelerate the rural out-migration and aggravate the farming situation in the mountains. Yak farming is becoming increasingly vulnerable as the population of younger rural people is reduced and productivity of grasslands declines. Although difficult to achieve, appropriate schemes are needed to make yak farming attractive to the mountain youth. The high altitude communities shall continue to benefit substantially from the *Cordyceps* business. However, it is of paramount importance to invent and promote suitable harvesting techniques to ensure minimum damage to the grassland. It is hoped that the future government interventions will focus on increasing the productivity of grasslands, improving the breed of yaks and regulating *Cordyceps* collection to avoid grassland degradation. The government also needs to put in place a levy mechanism, whereby a small percentage of income from *Cordyceps* is invested in grassland renovation. In essence, yak farming is at a crossroads where a firm decision is needed to either strengthen the age-old tradition or witness its gradual extinction.'

AUTHORS' CONTRIBUTIONS

Both authors contributed to development of the concept and survey questionnaire, field interviews and data collection, data analysis and drafting of the manuscript. The final manuscript was approved by both authors.

ACKNOWLEDGEMENTS

The authors gratefully acknowledge the financial support of the Royal Government of Bhutan. We also thank Mr. Thubten, Mr. Samten Nidup, Mr. Harilal Nirola and Mr. Karma Dorji for their valuable assistance during the field surveys.

REFERENCES

1. Biernacki P, Waldorf D: Snowball sampling. *Sociological Methods and Research* 1981, 10:141-163.
2. Boesi A, Cardi F: *Cordyceps sinensis* medicinal fungus: traditional use among tibetan people, harvesting techniques, and modern uses. *Herbalgram* 2009, 83:52-63.
3. Cai L, Wiener G: *The Yak*. Thailand Food and Agricultural Organization of the United Nations, Bangkok; 1995.
4. Cannon PF, Hywel-Jones NL, Maczey N, Norbu L, Tshitila TS, Lhundup P: Steps towards sustainable harvest of Ophio*cordyceps* sinensis in Bhutan. *Biodiversity Conservation* 2009, 18:2263-2281.
5. Gould R: Himalayan Viagra, Himalayan gold? *Cordyceps sinensis* brings new forces to the Bhutanese Himalaya. *Tropical Resource Bulletin* 2007, 26:63-69.
6. Jacoby J, Miichael M: Three-point Likert Scales are Good Enough. *Journal of Marketing Research* 1971, 8:495-500.
7. Joshi S, Shrestha RM, Sherpa D, Shrestha B, Lun Y, Dorji Y, Tingyu G, Jioubing N, lai J, Ciren Y, Manohara K, Ritu V: *Gender and Pastoralism in the Rangelands of the Hindu Kush Himalayas: From the margins to new heights*. International Centre for Integrated Mountain Development, Kathmandu, Nepal; 2013.
8. Khan, Obaidullah. 1995. Forward speech. In The yak, Bangkok, Thailand.: FAO (Food and Agricultural Organization of the United Nations) Regional Office for Asia and the Pacific.
9. Körner, Christian, and Masahiko Ohsawa. 2005. Mountain Systems. In Ecosystems and Human Well-being: Current State and Trends. Millennium Ecosystem Assessment, Washington, DC: Island Press 681–716.

10. Landau S, Everitt BS: *A Handbook of Statistical Analyses Using SPSS*. FL Chapman & Hall/CRC, Boca Raton; 2004.
11. Likert R: A Technique for the Measurement of Attitudes. *Archives of Psychology* 1932, 140:1-55.
12. Meerza SIA: Rural–urban Migration and Its Consequences on Rural Children: An Empirical Study. *Asian Social Science* 2010, 6:176-181.
13. Miller, RA. 2009. The Cordyceps sinensis medicinal mushroom. www.nexusmagazine.com/products/the-cordyceps-sinensis-medicinal-mushroom-detail. Accesed 20 September, 2014.
14. Mishra C, Samanta B, Tsewang N, Bhatnagar YV: Multiple Use of Trans-Himalayan Rangelands: Reconciling Human Livelihoods with Wildlife Conservation. In *Wild rangelands: Conserving Wildlife While Maintaining Livestock in Semi-arid Ecosystems*. Wiley-Blackwell, Chichester; 2010:291-311.
15. Panda AK, Swain KC: Traditional uses and medicinal potential of *Cordyceps sinensis* of Sikkim. *Journal of Ayurveda Integrated Medicine* 2011, 2:9-13.
16. Roder W, Gratzer G, Wangdi K: Cattle grazing in the conifer forests of Bhutan. *Mountain Research and Development* 2002, 22:1-7.
17. Shaha, BKP. 2000. Economics of yak farming with relation to tourism in Nepal The third ınternational congress on yak. Lhasa, China. Kenya: International Livestock Research Institute.
18. Shreshtha, Surendra. 1994. Evolution of mountain farming systems. Proceedings of the FAO/ICIMOD seminar at Lumle Agricultural Research Center. 13–29. Kathmandu, Nepal.
19. Stockdale A: Rural out-migration: community consequences and individual migrant experiences. *Sociologia Ruralis* 2004, 44:167-194.
20. Stone R: Last stand for the body snatcher of the Himalayas? *Science* 2008, 322:1182.
21. Tenzin, K. 2009. Current Harvesting and Management Practices for Yartsa guenboob in Bhutan Regional Conference on Cordyceps Management and Marketing Bumthang. Bhutan: Ministry of Agriculture.

22. Tulachan PM, Neupane A: *Livestock in Mixed Farming Systems of the Hindu Kush-Himalayas: Trends and Sustainability*. Nepal International Centre for Integrated Mountain Development, Kathmandu; 1999.
23. Wangchuk D: *Cordyceps sinensis: Is Domestication Possible?*. Bhutan's National Newspaper, In Kuensel; 2008.
24. Wangchuk K, Dorji K, Lhendup U: *Crushing the Bone: Minimizing Grazing Conflicts in Community Tsamdro- A Case Study from Dhur Village*. Choekhor Geog, Bumthang, Bhutan; 2006.
25. Wangchuk K, Wurzinger M, Darabant A, Gratzer G, Zollitsch W: The changing face of cattle raising and forest grazing in the Bhutan Himalaya. *Mountain Research and Development* 2014, 34:131-138.
26. Wangchuk K, Gyaltshen T, Yonten T, Nirola H, Tshering N: Shrubland or Pasture? Restoration of Degraded Meadows in the Mountains of Bhutan. *Mountain Research and Development* 2013, 33:161-169.
27. Wangchuk S, Nawang N, Sherub : *Impacts of Cordyceps Collection on Livelihoods and Alpine Ecosystems in Bhutan as Ascertained from Questionnaire Survey of Cordyceps Collectors*. UWICE Press, Bumthang Royal Government of Bhutan; 2012.
28. Wangmo S, Wangchuk K: The collectors' dilemma: Constraints and opportunities for community based *Cordyceps* management in Lingzhi. *Journal of Natural Resources Bhutan* 2008, 4:172-174.
29. Wiener, G.,Jianlin, H. and Ruijun, L. 2003. Breeding, Crossbreeding and Hybridizing in Yak. In Bangkok, Thailand, 33–35. Food and Agriculture Organization of the United Nations, Regional Office for Asia and the Pacific.
30. Winkler, Daniel. 2013. Steps towards Sustainable Harvest of Yartsa Gunbu (Caterpillar Fungus, Ophiocordyceps sinensis) Proceedings of the 7th International Medicinal Mushroom Conference. Beijing 635–644.

14

'It is Better to Herd than be Herded': Making a Living with Goats in the Bajío Region, Mexico

David Oseguera Montiel[1], Nícola Maria Keilbach Baer[2], Akke van der Zijpp[1], Chizu Sato[3], and Henk Udo[1]

[1]Animal Production Systems Group, Wageningen University, Wageningen 6700 AH, the Netherlands

[2]Colegio de Michoacán A.C., Martínez de Navarrete 505, Col. Las Fuentes, Zamora 59699, Michoacán, Mexico

[3]Sociology of Consumers and Households Group, Wageningen University, Wageningen 6700 DA, the Netherlands

ABSTRACT

Goats are renowned for their resilience in harsh environments and their relatively low investment for maintenance. Goat husbandry is thought to be a tool for poverty alleviation. Empirical evidence of this is scant. This research analysed the role of goat husbandry in supporting the livelihoods of smallholders from the Bajío region in Michoacán, Mexico. The Bajío is renowned for the good cropping potential of the land; smallholder goat husbandry is present too but largely unstudied by scholars and ignored by policy makers. The smallholders in the study area deploy a range of assets, natural, physical, social, human and financial, in goat husbandry. Their goat husbandry is dairy-oriented; it is a source of weekly income and insurance and therefore an alternative to out-migration. Farmers' relatively high social capital allows them to access cheap crop residues and take turns herding flocks. The goat dairy market is controlled by a powerful caramel industry. In turn, the margins smallholders obtain are rather limited. The nutritional value of goat milk is not exploited in their households as it is seen as a 'fever' cause, related to brucellosis. Qualitative and quantitative methodologies are based on the sustainable livelihoods approach linked to actor-oriented approaches. The study revealed smallholders' agency by engaging in goat husbandry to deal with a complex institutional and political context dominated by economic liberalization intertwined with local realities such as the agroecology. We emphasize the importance of these findings in development strategies for small-scale goat husbandry systems.

BACKGROUND

Goat husbandry is considered to have great potential to improve the livelihoods of poor people (Sinn et al. [1999]; Peacock [2005]; De Vries [2008]). Compared to cattle, goats are easier to raise in resource-poor households. This is because goats are resilient animals that can cope with relatively low quality feed and scarce water (Morand-Fehr [2005]). In the global South, goat husbandry produces more value than single-sided production-economic criteria. Goat husbandry plays a role in financial security, women's empowerment and insurance (Bosman et al. [1997]; Dossa et al. [2008]). Furthermore, goat husbandry exemplifies smallholders' agency, referring to smallholders' capacity to act and make choices. Smallholders' agency is often in response to an adverse context for smallholders, which in Latin America is partly the result of the neoliberal policies promoting free trade, privatization and deregulation, among others.

Under the neoliberal paradigm, Mexican smallholders are portrayed as backward or inefficient and hence the rhetoric that Mexico needs a 'modern' agricultural sector (Toledo [1992]). Smallholders are often deemed as the 'nonviable' (Bebbington [1999]) and therefore the way forward for neoliberal planners is to intensify and modernize the agricultural sector. For example, in 1992, the Mexican government, arguing that rural smallholders lacked productivity, launched the counter-land reforms that opened the door for seizing *'ejidos'* - smallholders' community-owned land, which was about half of the agricultural land until 1991. This policy resulted from the negotiations of the North American Free Trade Agreement (NAFTA) (Schmidt and Gruben [1992]). Furthermore, public funds for the smallholder sector, in the form of subsidies and extension services, were withdrawn, and BANRURAL (a farmers' bank) and CONASUPO (an institution to guarantee fair crop market prices) were dismantled. Under NAFTA, there has been an unprecedented growth of commercial food corporations with oligopolistic control of food commodities (Ochoa [2013]). This context for rural smallholders is further complicated by the unprecedented wave of violent crime often related to drug trafficking (Pereyra [2012]). And yet smallholders are still resisting this hostile context. On and off-farm diversification has been their main survival strategy in Latin American countries (Kay [2008]).

In Mexico, small-scale goat husbandry is one of these on-farm diversification strategies. Goats can be found across the country (INEGI [2007]). Although goat husbandry started during the Spanish colonization in the 16th century, little is known about the role of goat husbandry in contributing to smallholders' livelihoods. It might be a good option particularly among rural poor people to fight poverty which affects approximately half of the Mexican population (CONEVAL [2009]). This paper aims to study small-scale goat farmers' agency to make their living in a complex context, by using a sustainable livelihood framework that integrates actor-oriented approaches. We argue that small-scale goat husbandry is functional and well-adapted to the agroecological conditions of the Bajío region, a prosperous cropping area in central Mexico. We also unravel the threats and opportunities involved in small-scale goat husbandry as a livelihood strategy. In the following sections, we present the theoretical framework and methods and the historical background within which goat husbandry currently operates in the region. There follows an in-depth analysis of how goat husbandry has adapted to the local context and contributes to smallholders' livelihoods.

THEORETICAL FRAMEWORK

We used the sustainable livelihoods approach (SL) to analyse how smallholder goat farmers generate livelihoods in their specific agroecological and socio-economic context. A livelihood comprises: '[A]ssets…activities and the access to these (mediated by institutions and social relationships) that together determine the living gained by the individual or household' (Ellis[2000], p.10). Assets or capitals for smallholders are natural capital comprising communal land grazing, water and crop land; physical capital comprising tools, machinery, infrastructure (e.g. roads, railway) and livestock; social capital comprising reciprocity, associations, cooperation and trust; human capital comprising people's skills, knowledge, health, education and traditional knowledge transmitted through generations; financial capital comprising money stored in a bank or at home, and credits or loans (Chambers and Conway [1992]; DFID 1999; De Haan [2000]; Ellis[2000]).

Mapping out capitals according to different socio-economic strata of households can help to identify where pro-poor support can have impact (Bebbington [1999]). Socio-economic strata are often identified on how households themselves define being poor, sometimes in relation to their assets (Kristjanson et al. [2007]), however, households' capabilities can be important too in defining well-being by households. Capability refers to being able to eat well, dress, live without shame and have a social life among others (Chambers and Conway [1992]). Furthermore, it refers to being able to respond to shocks or stress, but also in being proactive by taking up opportunities to enhance their livelihoods, such as making use of information, collaborating with others, experimenting, and using new resources and services (Chambers and Conway [1992]). This study tries to understand the role of goats in terms of households' capitals and capabilities.

Capitals can be converted to other forms of capitals and there are tradeoffs on how capitals are used (DFID 1999). For example, goats, a natural capital, can be used to generate outputs like meat, milk and manure, which are often a source of income (Ellis [2000]). Goats are liquid assets to be used in times of cash need, so for smallholders, goats are an insurance (Bosman et al.[1997]). Goat management, however, can also lead to the destruction of other capitals. By way of example, in Indonesia, goat manure is a valuable output for cropping by enhancing land fertility, a natural capital, but at the same time, manure is piled up near settlements and pollutes ground water (Budisatria [2006]).

Access to capitals is central in understanding livelihoods while reflecting how policies and institutions affect access to capitals is also important (DFID 1999). In access to capitals, politics and power relations play a role. As such, 'a livelihood is organized in arenas of conflicting [and] co-operating actors' (De Haan and Zoomers [2005], p.34). Yet, politics have been overlooked in livelihood studies (Scoones [2009]). Politics relate to how macro policies (e.g. neoliberal globalization in Mexico) and power relations among actors affect the access to capitals. In the Mexican rural context, different actors often collide and intertwine (Long [1998]). Drawing on these concepts, we postulate that goat husbandry is a livelihood strategy adapted historically to its specific context and that goat-keeping households employ a range of capitals, the access to which is influenced by the context. This refers to the institutional and political context in which goat husbandry is embedded, such as the agroecology and everyday relations of farmers among themselves and with other actors.

STUDY AREA

Agroecological Conditions

The study was conducted in northern villages of Michoacán, a part of the Bajío region which is a relatively large area that encompasses territories of four central states (Michoacán, Guanajuato, Querétaro and Jalisco) of Mexico (Chávez-Torres [2005]). The region is characterized by plains interrupted by multiple hills and volcanoes. The altitude ranges from 1,000 to 2,000 m above sea level, rainfall average is 800 mm and most precipitation is during the summer period (June to October). Ambient temperature averages 20°C. The Bajío region is a river basin area that has a high crop production potential, with irrigation in some areas. Land is cultivated for maize, sorghum, wheat and now notably cash crops (vegetables and fruits) in greenhouse systems for export to the USA (Chambers et al. [2007]). Two main forms of cropping can be found: high-input industrial agriculture and traditional low-input crop production (González-Martínez [1992]), a pattern present since colonial times where irrigated wheat was the predominant crop (Chambers et al. [2007]). Smallholders cultivate maize, sorghum, wheat, chickpeas on rainfed marginal land as well as on high potential agricultural land, mainly for self-consumption (by humans and animals), but as cash crops too. Besides cropping, smallholders are also engaged in seasonal wage labour, livestock husbandry and temporal migration to the USA. The region has one of the highest rates of out-migration to the USA in the country (Arias and Mummert [1987]).

Historic Context of Goat Husbandry in the Region

Goats were introduced by the Spaniards in the 16th century during colonization. In this period, the goat population had an exponential growth, due to the abundance of feed sources, grazing land and crop residues (Braudel [1984]). Spaniards seized indigenous peoples' land, which became the '*Haciendas*' - huge farms. In the search for more crop land, Spaniards moved to the Bajío region, where goats were used mainly to clear off the vegetation for later land cultivation (Baroni-Boissonas [1990]). Then goats became the most prominent livestock species (Rabell [1986]). Goat meat was also a food source for mine workers and goat fat was used to make candles for the mines of the region. In the 18th and 19th centuries, when cropping became an important activity in the Bajío, cattle outnumbered goats. This was linked to the development of more sedentary farming systems (Rabell, [1986]). Goats were kept in extensive grazing systems and cared for by hacienda workers (Zendejas-Romero [2003]).

By the 1930s, some haciendas were dissolved and the land was distributed among smallholders, who were entitled to work a plot of land in a so-called *ejido* system. Currently in the Bajío region, small-scale goat husbandry based on semi-extensive grazing management is relatively popular, especially in villages of the Chapala lake basin area and the Lerma river. Goat milk is the main input for *cajeta* - similar to the English caramel. The *cajeta* industry plants are found in the region.

Smallholders keeping goats are colloquially known as *chiveros*. Goat meat has a local market for a popular dish called *birria*.

METHODOLOGY

The field work was conducted in two periods: (1) from February to July 2007 and (2) from May 2008 to July 2009. A snowball technique (Biernacki and Waldorf [1981]) was used to identify villages where small-scale goat husbandry was present. In the first period, villages of Tanhuato and Ecuandureo municipalities were visited by the field researcher. Semi-structured interviews gave a better understanding of goats' multifunctional roles and farmers' constraints. Smallholders herding their flocks were approached and appointments were made for interviewing. Secondary data was sought to understand the history of goat husbandry in the region. Two cross-sectional surveys were carried out in four villages, Cieneguitas, Los Charcos, San José de Vargas and Tinaja de Vargas from the municipality of Tanhuato (Figure 1). The first survey was used to characterize smallholders' livelihoods assets by their socio-economic strata. In this survey we also investigated how goat husbandry was valued by households; this included identification of the different functions represented in goat husbandry and small holders' other strategies. This study was approved by the Wageningen Institute of Animal Science, Wageningen, the Netherlands.

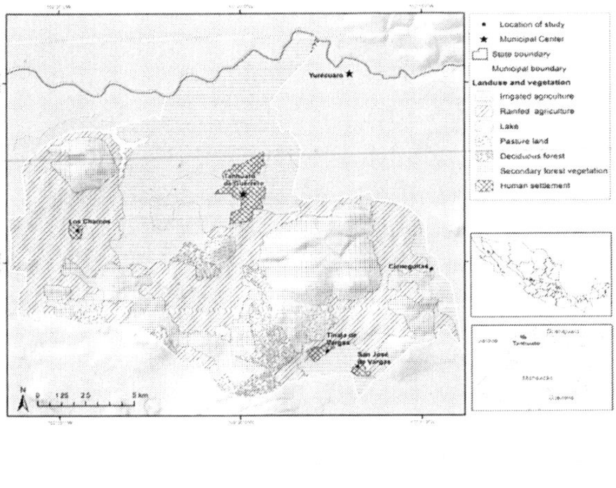

Figure 1: Study site. Vegetation and land use of the study area. Source: INEGI [2011].

Goat keepers were subdivided into three wealth groups: 'poor' , 'medium' and 'better-off'. In conjunction with an NGO employee who had been working directly with goat farmers to control brucellosis, a key informant was identified within each village to help sort fellow villagers into the wealth strata. From a census by the NGO, a list was obtained with names of the head of the household and his or her number of goats. The second survey aimed to understand the views of goat farmers' neighbours about goat husbandry. A random sample of 145 households was interviewed using a questionnaire with closed questions.

The second part of the study involved a longitudinal survey and qualitative methods. The purpose of the survey was to calculate the gross margins of goat husbandry and crop production. Data about inputs and outputs of crop and goat production over the year were obtained from 18 farmers. Farmers recorded the information in notebooks, which were then collected on four occasions throughout the year. Gross margins were calculated as the difference between the outputs, such as crops, milk, goat kids and the inputs, for example, fertilizers, sprays, feed, vaccines, wormers and antibiotics. The contribution of goat husbandry and cropping to cover a family's basic necessities (e.g. food, housing, health, education) was evaluated by comparing gross margins with the poverty threshold in rural Mexico which is 15,384 Mexican pesos (MX$) (1 MX$ = 0.08 USD, source: Banixco [2012]) per year per capita (CONEVAL [2009]). Milk price was also compared to inflation in Mexico from 2006 to 2008.

Qualitative methods involved ethnographic observations in households, milking sites, grazing areas, farmers' meetings and milk collection. We also used rural appraisal techniques such as group discussions about goat husbandry versus temporal migration to the USA (three group discussions), mapping and transects (three completed), semi-structured interviews with farmers ($n = 19$) and other stakeholders ($n = 10$), and informal talks among smallholders and stakeholders.

Quantitative data was described with R (R Core Team [2012]) and ggplot2 was used for graphs (Wickham [2009]). Interviews were generally audio-recorded, or else notes were taken. Audio-recorded interviews (ranging from 1 to 2 h) were fully transcribed in Spanish. Qualitative analysis was done by (1) coding material, (2) identifying themes and by (3) describing and exploring themes. We used Weft QDA for coding (Fenton [2006]).

RESULTS

Capital Status of Different Groups of Smallholder Farmers

Figure 2 shows the distribution of land, the main natural capital, in the four villages. Most of the cultivated land is used for growing maize and sorghum; the latter is more common in villages of San José de Vargas and Tinaja de Vargas (plot B). Some plots have irrigation systems which allow farmers to cultivate the land twice per year with a rainfed crop - often maize or sorghum - and an irrigated crop (often wheat) in the dry season. The communal non-cultivated land is mostly shrub, which is used mainly to graze goat flocks during the rainy season.

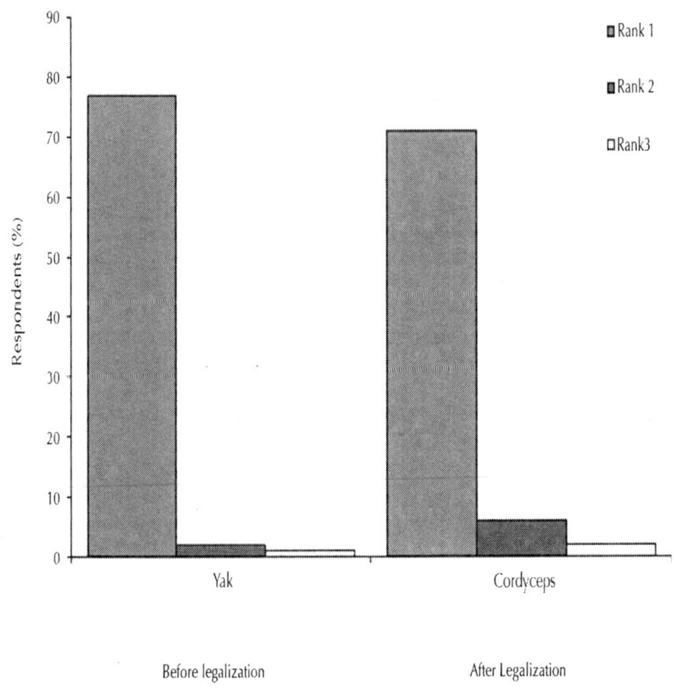

Figure 2: Land distribution in the villages (A) and grown crops (B). Each village has communal land, which is shrubland in the hills and parceled land, which is found in the plains. Source: RAN [2010].

Table 1 shows the distribution of some of the capitals across the wealth groups. Better-off goat keepers had on average 15 ha of cropland (first and third interquartiles (IQ) 11 to 21). Most of this land was of higher quality, located in the valleys, leveled and some plots had irrigation. Most of these farmers had large flocks, on average 131 goats (80 to 154 IQ). These farmers invariably had a truck and a tractor.

Table 1: Descriptive statistics of farmers' assets and household characteristics according to their strata

	Median	**IQ**	**Mean**	**SD**	***N***
Land (ha)					
Poor	2.0	1.0 to 2.5	2.0	1.2	14
Medium	5.0	4.0 to 7.0	5.8	2.6	21
Better-off	15.5	11.0 to 21.0	15.8	9.2	8
Goats (*n*)					
Poor	30	22 to 45	37	21	17
Medium	70	32 to 123	91	83	23
Better-off	107	80 to 154	131	87	8
Household size (*n*)					
Poor	5	3 to 6	5	2	17
Medium	5	4 to 6	5	2	23
Better-off	4	4 to 6	5	2	8
Age household head (years)					
Poor	49.5	44.5 to 54.0	50.0	12.1	16
Medium	51.0	39.0 to 64.0	50.2	13.7	23
Better-off	39.5	36.3 to 58.3	45.6	16.7	8

IQ, interquartile range; SD, standard deviation; N number of observations.

Oseguera Montiel et al.

Oseguera Montiel et al. Pastoralism 2014 4:9, doi:10.1186/s13570-014-0009-2

The poor and the medium wealth groups had less of the above capitals. Poor households had on average 37 goats (22 to 45 IQ). Two-thirds of the households in the poor group owned 2 ha of land on average (1 to 2.5 IQ). Poor households' crop land was often in communal areas and was known as *ecuaros*, which were plots of approximately 1 ha in the edges of the hills. The medium wealth group had on average 90 goats (32 to 123 IQ). They owned relatively better crop land than the poor and on average they owned 6 ha of land (4 to 7 IQ). There were however, landless farmers in these two groups too ($n = 3$ in poor and $n = 2$ in medium). Two-thirds of medium farmers had a truck and 15% had a tractor, whereas only a quarter of poor farmers had a truck and only one poor farmer had a tractor.

There were no differences among wealth groups in household size or age of the household head (Table 1). Despite the similar size of the households, health status, migration and age of children played a role determining the wealth stratum of the household. Households were classified as poor when the head of the household (a man) was unfit for physical work due to illness. For example, one household deemed as poor had three children in their teens, approximately 0.75 ha of crop land and 45 goats. The man, however, was incapable of doing any work as he was in a wheelchair. Relatively, young married couples with infants, absence of the man (head) due to migration to the USA and being hired as labourers were other reasons given to classify households in the poor group. Two-thirds of the households reported having more than 10 years of experience in goat husbandry, regardless of their socio-economic group.

There was a range of social capital forms among households. It was remarkable the companionship that derives in sharing responsibilities for herding goats. Some households also engaged in entrepreneurial activities such as growing alfalfa (one farmer with irrigated land and another with financial assets). In one of the villages, farmers constantly communicated to help each other during herding, often to find lost animals or share news, using walkie-talkies. The form of social capital also differed among the wealth groups. The medium and better-off groups for example made strong ties with some of the personnel from the NGOs. An NGO person could be invited to have lunch at their houses. In return, these households could have their flocks vaccinated first.

There was a history of rivalries between families of two villages which weakened social capital at community level. This rivalry had led to deaths in both villages. During the field study for example, farmers from one of the villages reported they felt oppressed by an extended family from the neighbouring village. These farmers were often insulted by this family. 'They also come at nights to dare our children' , the farmers reported. For some farmers, the situation was unbearable and they fled because of fear of violence against them or because of direct death threats. Two extended families migrated to nearby villages together with their goats, but were unable to harvest their crops any more. Three other households sold their flocks and tried to make their way elsewhere.

The wealth ranking exercise showed how goat farmers perceived their well-being. In general, farmers considered themselves to be better-off compared to those having no goats (physical capital). Figure 3 shows that goats were not just a source of income. Security was also important. Having goats was a guarantee of access to credits in their village. '[W]ith a small flock one can defend better than one who owns nothing...being a goat farmer you always have one or two *pesos*in your pocket'. 'You are what you have, if you have goats you can get credits, if you only have land you will not get them. [The logic of the lender is].. [a farmer] will not sell land to pay but he will sell goats' (farmer, San José de Vargas). In the group discussions, it was stated that having a flock was also a way to be your own boss and not to be fully dependent on being a hired labourer '*mas vale arrear que ser arreado*' (it is better to herd than be herded). Furthermore some job offers were not acceptable. Farmers reported that they were hired to apply a lot of chemicals in cropping (e.g. pesticides and herbicides), they have long workdays (15 hours) and wages were not good either. Farmers also benefit by having *birria* (the local goat meat dish) at their disposal. About 11% of the male goat kids were for self-consumption. Farmers could also please friends or relatives with a *birria* dish for special occasions such as in children's graduations from primary or secondary school. Goat male kids were also given as gifts; about 14% of the male goat kids were given to friends or to neighbours to thank them for allowing the goats to graze their crop residues.

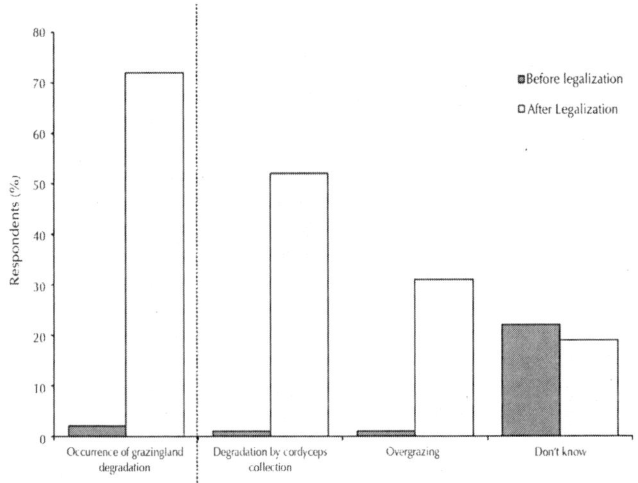

Figure 3: Goat husbandry functions based on cross-sectional survey of farmers' views ($n = 46$).

For some senior farmers, goat husbandry was the base for accumulating other capitals (i.e. land and cattle). The best example of this was a handful of extended family households who had managed to consolidate relatively large flocks ≥400 equal to about 30 goats per capita. Farmers reported that their parents had started with just a handful of goats. Among farmers who managed to consolidate large flocks, especially middle-aged farmers, migration to the USA was not considered an option. One said 'I am happier here with my goats' (farmer, San José de Vargas). Young men however, with a small flock (<15 goats) and little land of low crop production potential were relatively eager to migrate.

Livelihood Strategies Portfolio

Figure 4 shows that households engaged in a range of activities to optimize their cash income, food self-sufficiency and risk management. The range of activities tended to be wider for households in medium and better-off groups. Cropping was the most frequently reported activity across the three wealth strata of households. External inputs were fertilizers, seeds, sprays, hired labour, diesel, machine hiring and transportation. Therefore, households in particular farmers in medium and better-off groups sought to get credits to cultivate land. Households grew for the market and for home consumption. In general, only households in the better-off and medium groups could grow for the market (sorghum and maize). Land with less cropping potential, which was cultivated with fewer external inputs, was often used to grow maize for home consumption by poor households, which shows the importance of maize for food self-sufficiency in poor households. Households with irrigated land cultivated wheat and alfalfa. Chickpeas were also cultivated and used as fodder for goats. Farmers cultivated chickpeas in plots that have some residual humidity, and often shortly after, they noticed that their first crop (e.g. maize) would not give a good harvest when rains were delayed, which shows how farmers manage risks. Another risk management strategy was to cultivate sorghum rather than maize, because sorghum was perceived by farmers as more drought resistant than maize. Middle and better-off farmers cropped more land than they owned because they hired or borrowed crop land. In contrast, poor households could not always crop their own plots, because of lack of cash to get inputs or because of lack of male labour in the household.

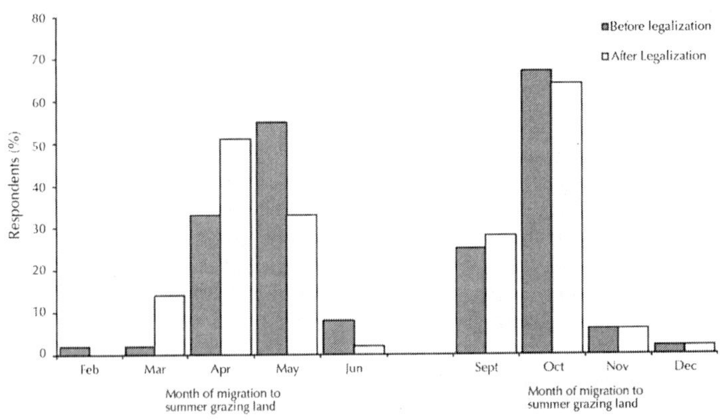

Figure 4: Household strategies portfolio based on cross-sectional survey of farmers' views, ($n = 46$).

Among households from lower and medium strata keeping one or two sows to sell piglets was common, especially in the village Los Charcos. Cattle keeping was negligible for households in the lower wealth strata compared to households in the medium and in the better-off groups. Better-off households keeping cattle owned relatively large cattle herds of about 100 head, whereas the few poor households with cattle had only one or two animals.

Household income was often complemented with remittances from the USA, agricultural wages and non-agricultural salaries from the nearby towns (i.e. Zamora and Tanhuato). Farmers reported that remittances were used to get inputs for the flock such as feed.

Women were engaged in specific income-generating activities. Women from the medium group were selling chicken parts and making cheese. Making prayer beads and other religious crafts by sewing was common among women from poor households. Women in the poor and medium groups were also engaged in seasonal work, such as harvesting vegetables and fruits, and in permanent jobs, such as packing strawberries for the frozen fruit industry of Zamora.

Goat Husbandry as Pastoralism

In general, goat husbandry in the Bajío region can be characterized as a pastoral activity. Farmers were knowledgeable about the grazing behaviour of goats and their adaptation to the environment. Goats were herded to graze crops, crop residues and native vegetation in communal land and roadsides. Farmers herded their goats to graze native grasses, fodder trees and shrubs in the hills of communal land and roadsides during the rainy season (July to October), which is a period when goats were dried off from milking (Figure 5). The dry season lasts from mid-October to mid-June. In the dry season, goats were herded to graze crop residues of rainfed crop of maize, sorghum, chickpeas and irrigated crops of wheat. Farmers also herd their flocks to directly graze crops, mostly sorghum and chickpeas. Some other farmers kept their harvest to feed their goats in periods of poor forage availability. There were not many maize and sorghum residues left by the end of March. At this time of the year, most farmers stopped the grazing system and changed to stall-feeding, based on stored crop residues, crops and purchased forage and concentrates. Some farmers continued with grazing throughout the dry season and brought their goats to the communal land to browse shrub pods (*huizaches*), chickpeas and wheat residues. All farmers, however, had to supplement their goats' feed with concentrates or crop residues. Lactation started in mid-October and the peak of the production was around mid-April. The breeding season started in May when farmers let their bucks mate the does. Milk yields started dropping in July and August (Figure 5).

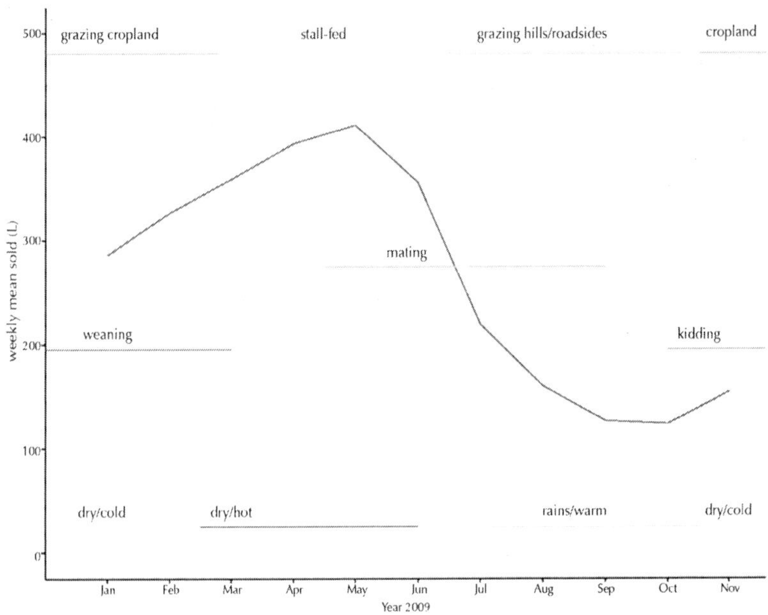

Figure 5: Lactation curve of goats, seasonal management and weather. Source: data on milk production and management from five farmers' records from villages of Michoacán.

Access to Land

Figure 6 shows a panoramic view of the valley during the rainy season. From the interviews, we learned that the access to natural capital land was becoming difficult. The land counter-reforms of 1992 that established individual property over *ejido* land affected the way villagers organized their cropping and livestock husbandry. The first problem due to this land tenure change was that there were less crop residues available for goats. Formerly, after the harvest, all crop residues were left for all village livestock to graze freely. 'Now everything has an owner' (farmer, San José de Vargas). Hence, crop residues had a price. Payments in kind (goat kids) were common to allow access to neighbours' crop residues. Farmers might let their goats graze neighbours' crop residues without asking permission. Goat farmers themselves recognized that there was some stigma for 'stealing' crop residues. Furthermore, the land counter-reforms of 1992 allowed external users to lease and purchase land. For example, in San José de Vargas, there was a mine in operation, which occupied a large portion of the communal land. In the vicinity of Los Charcos, a feedlot for up to 10,000 cattle was established. Los Charcos farmers were concerned about the water reservoirs as the feedlot uses water for the animals and also for slaughtering and meat packing. Added to this, the feedlot sewage was discharged in the village canal, and the farmers were very annoyed by pestilent clouds of dust that covered their village since the feedlot started to operate.

Figure 6: Grazing goats in shrubland. Herding to hills takes place during the rainy season in the study region; white patches in the background are greenhouses (photo by N. Keilbach).

Dairy Goat Farming: Markets and Margins

Goats are an important financial capital. Flocks were mostly crossbred goats of dairy breeds, such as Saanen, French Alpine and Toggenburg. The longitudinal survey showed that annual milk yields per head averaged 422 L (Standard deviation (SD) 106). Milk was sold at farm gate to the *cajeta*industry. Male goat kids were also sold. They fetch prices of around 270 Mexican pesos (MX$) (USD 22). Goat kid meat had a seasonal market at the end of the year (Christmas and New Year celebrations) coinciding also with the return of migrants. The role of goat husbandry as a source of income is shown in Figure 7. Plot A shows the importance of goat milk sales to the total on-farm gross margins (i.e. cropping and goat husbandry). Lower gross margins were obtained by households with small flocks, which belonged to the poor. There was a linear relationship between milk sales and gross margins. Plot B shows that cropping had a lower direct impact on on-farm gross margins than goats. To some extent, goat husbandry had a higher impact on gross margins when compared to cropping (plots A and B) because farmers used their own crops as feed, inputs for cropping were costly and crops failed due to droughts.

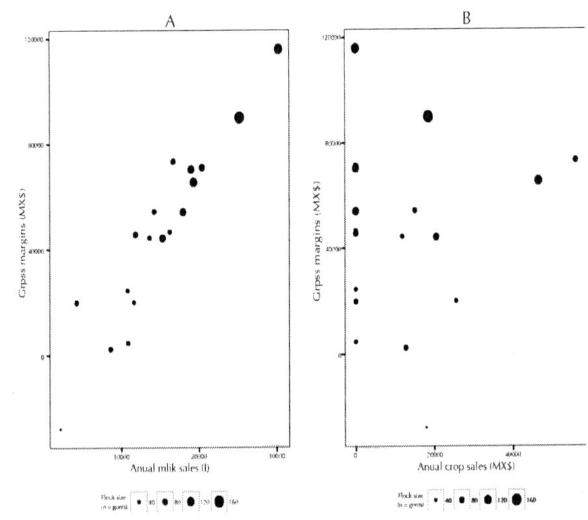

Figure 7: On-farm gross margins per year with regard to flock sizes. (A) On-farm gross margins according to annual farm milk production and flock size. (B) On-farm gross margins according to crop sales and flock size (n = 18).

Households in the medium and better-off strata obtained larger gross margins when compared to the poor, as shown in Figure 8 (plots A and B). The contribution of goat husbandry and cropping in relation to the poverty line is shown in plot B. The gross margins per capita of poor farmers (median = 4,987, IQ = 3,895 to 11,948) and medium farmers (median = 9,029, IQ = 7,736 to 11,723) were under the poverty line. Only a better-off farmer was well above the poverty line.

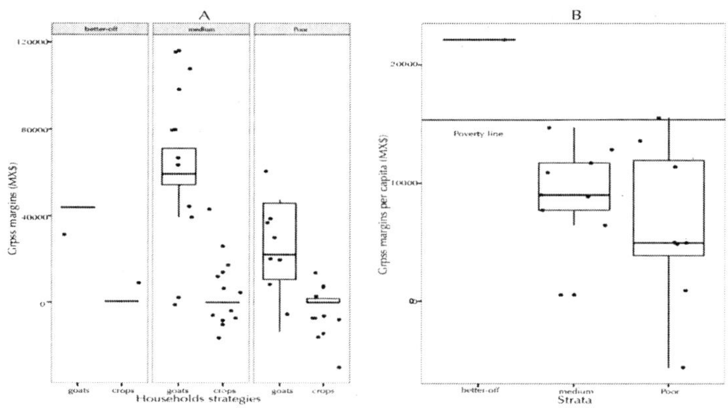

Figure 8: Gross margins from goat husbandry and cropping. (A)Box whiskers plots of gross margins per activity, dots are the observations. (B) Box whiskers plots of gross margins per capita in relation to the poverty line (n = 18).

The positive gross margins from goat husbandry were a reason for its popularity. For example, in one of the villages (Cieneguitas) where there were formerly three to four goat farmers, there were now 28 households with goat flocks averaging 40 head (SD 36), according to a census done by an NGO. '[T]here are a lot of goats now, the truck used to come for 600 litres, now two trucks leave full of milk ... now everybody has some goats' (farmer, Cieneguitas). Another farmer comparing cattle with goats said that 'cows do not produce, it takes two years before you can sell a calf' (farmer, San José de Vargas).

Farmers' main concern was the milk price though; they received $MX 4 (USD 0.33) per litre and they felt that the price did not increase at the same rate as input costs. Figure 9 shows how the milk price (adjusted to 2008 prices) fell during 2006 and 2007. In November 2007, the price increased until March 2008 and then started to fall again. The *cajeta* companies justified their farm gate milk price because farmers produced little milk and milk quality was low. The managers of the *cajeta* plants claimed that 'they [farmers] produced very little [individually], to fetch the milk at farm gate is very costly… they have very very inefficient systems, they keep many non-productive goats' (managers of the *cajeta* plants). The leading company in collecting the milk was a multinational company *Coronado*. There were two other prominent milk companies - *Cajeta Cabadas* and *Real the Potosí*. The *Coronado* plant processed over 12 million litres in 2010. We were informed that there was a 38% increase in milk processed in 2010 compared to 2008.

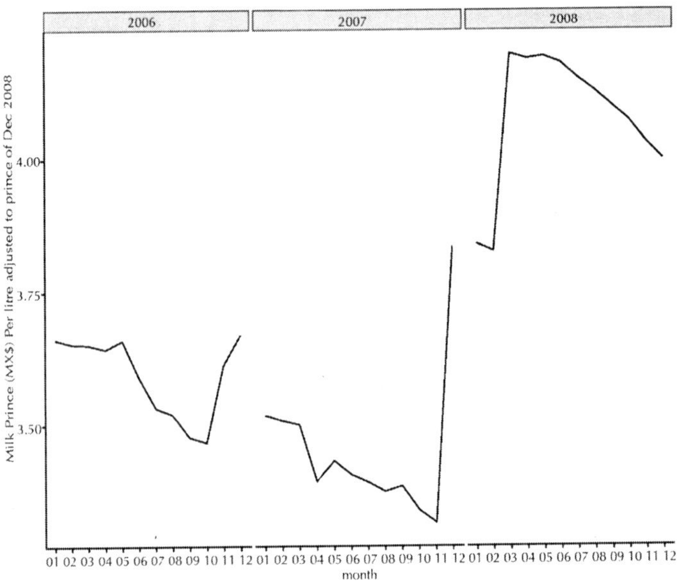

Figure 9: Goat milk price (years 2006, 2007 and 2008). Source: a farmer's records of milk sales.

The *cajeta* industry controlled the goat milk market. The managers of *cajeta* local plants were known as the '*patrón*' (the boss) by farmers. To prevent the farmers teaming up to demand a higher milk price, the industry paid a slightly higher price to farmers with larger flocks than to the majority of the farmers, which was a kind of *divide et impera* strategy. Adding water to milk was sometimes how farmers took revenge for the low price for their milk. They however risked paying a penalty because random samples were taken to detect diluted milk. There were also patronage strategies used by the industry to ensure that a farmer's milk production was sold to them. When goats were dry, farmers asked for credit from milk traders and the industry. In turn, farmers sold their milk to their credit providers. Credits were given without interest rates and were paid back gradually when the milk production was peaking again. Usually these credits were used for daily living expenses and were equivalent to one or two weeks of a household's milk production. If credits were not given to farmers, the industry risked losing their milk supply, because farmers then sold their milk to a competitor.

Selling milk to the industry at relatively low prices was not the only stressor for farmers. Brucellosis, a zoonotic disease (that can be transmitted from goats to humans), was endemic in the region (NGO personnel, personal communication). Milk processing to produce *cajeta* eliminates the risk of brucellosis for the consumers. The industry interviewees reported that there were no plans to pay premium prices for milk from brucellosis-free flocks. In 56% of the households surveyed, respondents reported having at least one family member who had contracted brucellosis. Therefore, as a preventive measure, households avoided goat milk consumption. Farmers reported that physicians recommend staying away from dairy goat products as it is a cause of 'fever'.

Labour and Knowledge

Goat husbandry involved only family labour. Men took decisions and received the cash from sales of milk. Women took the lead if their husband was ill, or when their husband migrated to the USA. Invariably, men were the herders and daily herding took 6 to 8 hours. If husbands were absent due to migration, boys (some just over 10 years old) could take over the main goat husbandry tasks such as herding. Milking was done by hand once a day in the morning. Women swept the pens after the men left for herding. Usually, one farmer alone was able to milk a flock of up to 70 head. For larger flocks, other family members came to help, such as the *chivero* wife (Figure 10). Women also helped in washing milk containers, raising goat kids and preparing the menfolk's lunch for herding. In extended households, women and young children helped in other activities like herding and giving medical treatments.

Figure 10: Woman milking a goat flock. Hand milking is predominant even when flocks are large. Tinaja de Vargas, Michoacán (photo by N. Keillbach).

Table 2 shows neighbours' opinions about goat husbandry. Neighbours preferred goat meat over goat dairy products. Goat farmers were seen as *'gente de trabajo'* (working people) by their neighbours, which was a compliment. In general, farmers received relatively high esteem among their neighbours. For some farmers, goat husbandry was an apprenticeship for their children. In the villages, the highest school education available was secondary education. Therefore, goat husbandry was attractive for children who could not attend school beyond primary school.

Table 2: The opinions of farmers' neighbours about goat husbandry ($n = 145$)

Villages				
	Cineguitas	**Los Charcos**	**San José de Vargas**	**Tinaja de Vargas**
Number of respondents (*n*)	37	59	27	22
Low esteem to goat farmers (%)				
Destructive	3	4	15	0
Odour	0	15	4	0
High and neutral esteem to goat farmers (%)				
Working people	16	17	11	14
Generate employment	3	9	18	45
Good but no specific reason	8	2	4	0
Neutral	70	53	48	41
Dislikes about goats (%)				
Nothing	68	37	56	77
Smell	27	43	22	23
Smell combined with flies	5	20	22	0
Likes about goats (%)				
Nothing	11	13	19	14
Dairy product	5	5	0	5
Meat	84	70	74	73
Meat and dairy	0	12	7	9

Oseguera Montiel et al.

Oseguera Montiel et al. Pastoralism 2014 4:9, doi:10.1186/s13570-014-0009-2

During transect walks; we observed that farmers were skilful in various aspects of goat husbandry such as herding and curing diseases. Farmers used different calls to herd their flocks, such as a call to urge the flock to come back, one to move on or a call to scare a coyote. Farmers also trained village dog pups to become herder dogs to protect flocks against thieves and coyotes. They predicted how long crop residues could last for their flocks and related the quality of crop residues to milk yields, as well as being very familiar with the properties of local vegetation (e.g. toxicity and nutritional value). Their knowledge of the local ecology was also of key importance. Accordingly, they planned routes to herd their flocks to the best spots for feeding. When grazing forage is decaying at the end of the rainy season and scant at the end of dry season, farmers feed their goats homemade rations of grains, forage and concentrates, illustrating farmers' knowledge on how to adapt feeding management according to the goats' needs. A common daily ratio per goat was 0.5 kg of commercial concentrates, maize or sorghum and 0.5 kg of maize stover or alfalfa. Farmers used natural and local remedies to heal udder cuts and resolve placenta retention. They were also familiar in how to use modern drugs for deworming, vaccinating and curing common infections, such as mastitis, pneumonia and diarrhea. They even knew how to treat acidosis in goat kids, a syndrome known locally as '*borrachito*' (drunk syndrome in goat kids), with oral rehydration salts for human use. But farmers perceived that their knowledge was not valued by outsiders. Farmers reported that an employee of the agricultural secretariat referred to them as ignorant and stinky.

Access Capitals Mediated by Institutions

There was one NGO 'Subcomité de Productores de Ovicaprinos del Estado de Michoacán' (SPOEM) working with goat farmers. This is an exceptional situation with regard to supportive institutions for goat farmers. SPOEM was not initiated within the villages but in the state capital, by an agronomist who envisaged the potential of smallholder farming systems and who intended to go in business with the most prominent goat farmers. The ultimate goal was to produce yoghurt (NGO personnel, personal communication). The entry point to start working with farmers was brucellosis control. The NGO was the channel through which governmental financial support for brucellosis control was given to goat farmers. The activities for brucellosis control included vaccination and testing to detect seropositive goats.

The ambitious brucellosis control programme (free for farmers) seemed to awaken interest from farmers to start goat husbandry and form groups. In some villages, prominent farmers were encouraged to form a group of 10 to 20 farmers and start an extension group

GGAVATT (Grupos Ganaderos de Validación y Transferencia de la Tecnología; Livestock farmers groups for technology validation and transfer). Farmers in these groups requested credit to acquire physical capital (e.g. pens).

We were informed that 48 farmers had received credit. We noticed that credit reached the medium and better-off farmers. The project better known as the '*tejabanes project*' (the pens project) was a kind of 'one size fits all' package of a pen and a milking machine per farmer, a cooling tank per group and also pasteurizers for some farmers. However, many things went wrong. First, pens were stall-fed prototypes, although forage grazing was the main management system. Second, farmers reported that pens were constructed with cheap low-quality materials, differing to the specifications in the original credit plan. One farmer reported having lost his goats when they got trapped and died within the post of the feeding fences, due to the bad design. Doors and fences of the pens were falling apart at the time of the field work. In addition, some better-off farmers already had a good size pen so they ended up having two pens. As a result, some of these pens were later used for other purposes. Figure 11 shows a 'modern' pen acquired through this project which was used to keep fighting cocks instead of goats, while some other farmers used the pens as storage for feed. Other equipment given was obsolete too. The milking machines were rarely used because they ran on petrol or electricity, so they required an extra input. 'We are not paid more for using it so it is not worth it' (farmer, Las Fuentes). The small ramp to milk goats served instead as a bench and table to have lunch, and was not used to milk goats. The cooling tank was placed where there was no electricity (i.e. in Tinaja de Vargas).

Figure 11: A 'modern' goat pen is used to keep fighting cocks.Michoacán, Mexico (photo by D. Oseguera).

Farmers launched a formal accusation of corruption due to the poor quality of the pens and some farmers stopped repaying their credit. In response, the government ceased financial support for the brucellosis campaign when these issues were brought up. Apparently, this was revenge against the NGO for supporting farmers in complaining about governmental corruption. Finally, some farmers who started as a GGAVATT group complained about not receiving any financial support compared to the neighbour village groups. Interviewed farmers reported that they had invested time and money in this group and nothing came of it. In summary, the pens project in Michoacán brought only problems, as one farmer reported.

DISCUSSION

Goat husbandry is part of the portfolio of smallholders' activities. Diversification has also been described in small-scale goat husbandry in northern parts of the country (Mora-Ledesma [2011]). The increasing number of households involved in goat keeping in various villages indicates that there is a growing interest among smallholders. Goat milk has become an important commercial commodity, while crops are mainly used for home consumption and as feed for the goats. Commercial cropping has become a risky activity for smallholders, whereas goat husbandry is relatively more feasible, especially for those having little or no land. As sale of milk gives a regular income, farmers try to optimize their income sources and manage risks through goat husbandry. This shows also farmers' agency; even among the poor, taking an active role to make their living a sustainable livelihood as they pursue autonomy and self-sufficiency. Goat husbandry seems to support those who argue that smallholders should not disappear, the so-called *campesinistas*(peasantists) (Kay [2008]).

We have shown that smallholders have a rich knowledge on keeping goats; this is very different to how industry managers and bureaucrats portray a *chivero*. Mastering a pastoral system in a relatively harsh environment is something to be recognized (Krätli and Schareika [2010]). This type of knowledge is described in Africa (e.g. Oba [2012]) and in other parts of Mexico (Mora-Ledesma [2011]). There is a rich goat husbandry knowledge transmitted through generations; part of this traditional knowledge originated during the Spanish colonial period around 500 years ago. The smallholder farming systems have subsisted all these years and are quite efficient, given the small size of cropland properties. Smallholders make use of the abundantly available natural capital, the so-called 'unproductive' shrub land.

Social capital is a key factor in goat husbandry, represented in community companionship, trust and the family members' work on various tasks of keeping a flock. A main drawback for smallholder goat farming was a weak community social capital due to violent events. Violence is increasing in the whole country, and in Michoacán, it is especially disturbing as murders were 100% higher in 2009 than in 2006 (INEGI [2014]). Although the analysis of this violent environment goes beyond the scope of this paper, it is an example of the consequences of livelihood destruction in Mexico's rural villages. As we were informed, farmers flee and stop farming, in order to avoid violence in one of the villages.

Our aim was also to understand the role of goats in improving poor people's livelihoods. This was done by investigating the role of goat husbandry among three groups of farmers defined as 'poor' , 'medium' and 'better-off'. Farmers said that in general, having goats was better than not having goats. For poor households, goat husbandry was more vital than for the medium and better-off. The last two groups had a wider range of activities. However, the role of goat husbandry to overcome the poverty line (i.e. fulfill basic necessities) is far from ideal. Most households were not earning enough per capita to overcome the poverty line of MX$ 15,348 per year (the equivalent of approximately US$ 1,250). Flock size is a factor in the overall on-farm gross margins (i.e. crops and goats). Poor farmers would need to own at least 30 adult dairy goats per capita to move to the medium group. We met cases of young couples in the poor category with relatively small flocks (approximately 15) where men were eager to migrate to the USA. Farmers in middle and better-off groups cultivated more land and of higher crop potential. Therefore they have more feed for their flocks. The poor have to restrain their flock size because they lack their own feed sources and have to buy extra feed. Therefore, claiming that goats can let people step out of poverty is not as straightforward as suggested in some articles (Peacock [2005]; De Vries [2008]).

It would have been useful to have more precise information about other cash resources, i.e. remittances and off-farm jobs. But it was very difficult to gain farmers' trust in mentioning money flows. Farmers found it rather strange that an outsider wanted to investigate their livelihoods. Farmers also found it hard to understand the overall benefits of our research, which were not tangible and were to some extent long term, so we did not want to risk stretching their confidence to the limit.

We did also meet two extended households that managed to keep relatively large flocks ≥400 goats (33 goats per capita), which allow them to acquire land. There were at least three key capitals for these extended families in their process of consolidation. First, there was a good individual social capital (i.e. family cooperation), second, human capital (i.e. labour of two or three generations including women and children) and natural capital (i.e. shrub land and village crop land residues). Poor households owned small plots (1 to 2.5 ha) and due to lack of financial resources and male labour because of USA migration, they could not cultivate their plots. Farmers with little crop land can however have access to crop residues of neighbours and access to communal grazing land to feed their flocks. But the land counter-reforms of 1992 threaten the access to these resources.

The neoliberal administrations tend to favour large farm operations (e.g. a feedlot) and mining. This is leading to resource exploitation of vast land areas by powerful companies. The potential of new conflicts due to delimiting access and creating competition for resources is just around the corner, as occurs elsewhere (Hollander [2013]). An example in Mexico is the experience with mines: local communities do not get what they are promised, while mines restrict access to communal grazing land and also pollute air, soil and water (Rodríguez Wallenius [2011]). Similarly, the feedlot enterprise in the vicinity of one of the villages is polluting air and water sources.

Goat milk is a commodity that generates a regular income flow for households; however, the milk trade is disadvantageous for farmers. A similar disadvantageous dairy marketing has been described in northern parts of the country (Gómez-Ruiz et al. [2012]). There was a huge contrast between the farmers' uneasy economic situation and the prosperous *cajeta* industry. Coronado is a subsidiary of *Bimbo,* worldwide the fourth largest food company and the largest bread manufacturer (Ochoa [2013]). Bimbo reported 20% larger profits in 2005 than in 2004.[a] Farmers were paid about 13% of the shelf price.[b] The caramel industry is the main winner here and as such it can be called a 'food empire' (Ploeg [2010]), which sets its rules such as milk price and the quality standards, e.g. checking watered milk, that are important for *cajeta* production.

[a]EMBITDA (earnings before interest, taxes, depreciation and amortization) were MX$ 7,191 million (Bimbo [2005]).

[b]A jar of 660 g of *cajeta* Coronado was MX$ 61 (PROFECO [2012]). *Cajeta* main inputs are milk and sugar; for 660 g of cajeta about 2 L of milk and 660 g of sugar are needed (employee, personal communication).

Farmers' main concern is the milk price which is stagnant in relation to inflation and to prices of their inputs. The goat milk market is in a vicious cycle where milk price is low and therefore the milk hygiene quality is low. Brucellosis, a zoonosis endemic in goat flocks of the region (Oseguera Montiel et al. [2013]), does not receive enough attention. This is not an issue in the eyes of the*cajeta* industry. Given the current circumstances, the risk of getting brucellosis is carried only by farmers' families. This is detrimental to farmers' livelihoods because affected individuals are not able to work and may develop permanent disabilities, e.g. arthritis, spondylitis (Corbel [2006]). Furthermore, brucellosis in goats is responsible for losses due to abortion and hence milk production is reduced (Corbel [2006]).

Controlling brucellosis could be an opportunity for farmers to find a niche market for a high-quality dairy product. Currently, such a market is exploited by a relatively small group of goat farmers approximately 20 km from the neighbour state of Guanajuato. Farmers might need to team up to achieve a better market. Stories of smallholder crop farmers forming cooperatives can be found elsewhere (King et al. [2012]). A cooperative led by women may be a way to empower women. Women are mostly involved in milking and cleaning corrals, but they could be playing a key role when goat milk is further processed (e.g. cheese making). Poorer households should be included in such plans too. Farmers' social capital such as cooperation and trust can be a starting point for developing cooperatives with the help of NGOs or governmental institutions. Unfortunately, good institutional support is lacking. Credit given to acquire 'modern' equipment in a 'one size fits all' package indicates the governmental institutions' desire for a modern agriculture sector. However, there was a mismatch between the package designed and the extensive grazing system used by the majority of goat farmers. A programme aiming to improve the pens of all farmers (not only the 'better-off') would have had a higher impact than the '*Proyecto de los tejabanes*'. The project only reached a really small fraction of the goat farmers of the region. In the whole state, there are 11,281 goat farmers (INEGI [2007]).

CONCLUSIONS

Goat husbandry in the Bajío, Mexico, is embedded in a complex context influenced by neoliberal policies that do not favour the small-scale farming sector. However, goat husbandry has a growing interest among farmers, partly because cropping has become risky and less profitable. Farmers see goats as a source of income, security, credit, prestige, independence, food, manure and apprenticeship for young children. The interest is present among all socio-economic strata. For the poor, goat husbandry was one of the main livelihood strategies. Better-off and medium group households had a wider range of activities. Wealthier farmers had relatively larger flocks and higher gross margins from goat husbandry than poor households. There were households who strengthened other capitals and strategies through goat husbandry. But these are just a handful of stories. The potential of goat husbandry as a tool for poverty alleviation is not visible yet. A dairy market oligopoly is a main drawback. This is partly linked to brucellosis in flocks, because the industry does not pay for high-quality milk; hence, there is no interest in tackling the brucellosis problem. Farmers are powerless against the dairy industry. There are opportunities for a better dairy market if brucellosis could be eradicated from the flocks. This could also reduce the risk of brucellosis in humans. Natural capital (i.e. communal grazing land) is key in goat husbandry. Historically, goat husbandry has persisted because of the abundance of this 'unproductive' land. Powerful external organizations have interests in this land and are therefore a threat for smallholder goat husbandry. Given the relatively low amount of crop land available for each household, we have shown that small-scale goat husbandry is productive, in contrast to the dominant discourse that smallholder goat husbandry systems are 'unproductive'.

AUTHORS' CONTRIBUTIONS

DOM conceived the study, carried out the field work, analysed the data and drafted the manuscript. NMK assessed the field work, HK, AJZ, NMK and CS commented on the manuscript. All authors read and approved the final manuscript.

ACKNOWLEDGEMENTS

Foundation Alfa and Omega in the Netherlands and The National Council for Science and Technology in Mexico (CONACYT) sponsored the first author and generously funded this research. We thank all farmers for their participation and hospitality during the field work. The authors wish to thank Dr. Anne Pearson for editing the manuscript and for valuable suggestions and comments with content of the manuscript. We also wish to thank Marco Antonio Hernández Andrade for his help in producing the map (Figure 1). Finally, we are thankful for the comments of an anonymous reviewer and to Carol Kerven for kindly editing the manuscript. The usual disclaimers apply.

REFERENCES

1. Arias, P, and G Mummert. 1987. Familia, mercados de trabajo y migración en el centro-occidente de México. Nueva Antropología. Revista de Ciencias Sociales 9: 105–128.
2. [http://www.banxico.org.mx] Banixco. 2012. Banco de México: Serie histórica del tipo de cambio peso-dólar. Accessed 27 August 2012.
3. Baroni-Boissonas, A. 1990. La formación de la estructuraagraria en el Bajío colonial. Siglos XVI y XVII. SEP Setentas, Ediciones. Centro de Investigaciones y EstudiosSuperiores en Antropología Social, México, D.F.
4. Bebbington, A. 1999. Capitals and capabilities: a framework for analyzing peasant viability, rural livelihoods and poverty. World Development 27: 2021–2044. doi:10.1016/S0305-750X(99)00104-7
5. Biernacki, P, and D Waldorf. 1981. Snowball sampling. Sociological Methods and Research 10: 141–163. doi:10.1177/004912418101000205
6. [http://www.mzweb.com.br/grupobimbo/web/arquivos/GrupoBimbo_II_AR_2005_eng.pdf] Bimbo. 2005. Bimbo 60 years, still growing heatlhier and stronger. Annual report 2005. Accessed 15 01 2014.
7. Bosman, HG, HAJ Moll, and HMJ Udo. 1997. Measuring and interpreting the benefits of goat keeping in tropical farm systems. Agricultural Systems 53: 349–372. doi:10.1016/S0308-521X(96)00047-9
8. Braudel, F. 1984. Civilización material, economía y capitalismo, siglos XV-XVIII. Alianza, Madrid.
9. Budisatria, IGS. 2006. Dynamics of small ruminant development in Central Java-Indonesia. Wageningen University, Wageningen.
10. [https://www.ids.ac.uk/files/Dp296.pdf] Chambers, R, and GR Conway. 1992. Sustainable rural livelihoods: practical concepts for the 21st century. Technical report, Institute of Development Studies
11. Chambers, KJ, SB Brush, MN Grote, and P Gepts. 2007. Describing maize (*Zea mays L.*) landrace persistence in the

Bajío of Mexico: a survey of 1940s and 1950s collection locations. Economic Botany 6 (1): 60–72. doi:10.1663/0013-0001(2007)61[60:DMZMLL]2.0.CO;2

12. Chávez-Torres, M. 2005. Bajío, Bajíos: Unidadesterritoriales de unaregión. Regiones: revistainterdisciplinaria en estudiosregionales 15: 117–140.
13. [http://www.coneval.gob.mx/Paginas/principal.aspx] CONEVAL. 2009. Consejonacional para la evaluación de la política de desarrollo social– reportaconevalcifra de pobrezaporingresos 2008 Accessed 14 January 2014.
14. Corbel, MJ. 2006. Brucellosis in humans and animals. World Health Organization, Geneva.
15. de Haan, L. 2000. Globalization, localization and sustainable livelihood. Sociologia Ruralis 40: 339–365. doi:10.1111/1467-9523.00152
16. de Haan, L, and A Zoomers. 2005. Exploring the frontier of livelihoods research. Development and Change 36 (27–47): 2005. doi:10.1111/j.0012-155X.2005.00401.x
17. de Vries, J. 2008. Goats for the poor: some keys to successful promotion of goat production among the poor. Small Ruminant Research 77: 221–224. doi:10.1016/j.smallrumres.2008.03.006
18. [http://www.eldis.org/go/home&id=41731&type=Document#.U7BqQ_mSySp] DFID. 1999. Sustainable livelihoods guidance sheets. Technical report, Department for International Development Accessed 21 May-2012.
19. Dossa, LH, B Rischkowsky, R Birner, and C Wollny. 2008. Socio-economic determinants of keeping goats and sheep by rural people in southern Benin. Agriculture and Human Values 25: 581–592. doi:10.1007/s10460-008-9138-9
20. Ellis, F. 2000. Rural livelihoods and diversity in developing countries. Oxford University Press, New York.
21. [http://www.pressure.to/qda/] Fenton, A. 2006. Weft QDA (version 1.0.1) computer software
22. Gómez-Ruiz, WJ, JM Pinos-Rodríguez, JR Aguirre-Rivera, and JC García-López. 2012. Analysis of a goat milk cheese industry in a desert rangeland of Mexico. Pastoralism: Research, Policy and Practice 2: 1–11. BioMed Central Full Text

23. González-Martínez, L. 1992. Respuesta Campesina a la Revolución Verde en el Bajío. Universidad Iberoamericana, México D.F..

24. [http://www.thebrokeronline.eu] Hollander, S. 2013. The tragedy of the deprived. Accessed 7 November 2013.

25. [http://www.inegi.org.mx] INEGI. 2007. Censo agrícola, ganadero y forestal. Accessed 16 January 2014.

26. [http://www.inegi.org.mx/geo/contenidos/recnat/usosuelo/inf_e1m.aspx] INEGI. 2011. Conjunto de datos vectoriales de uso de suelo y vegetaciónAccessed 30 May 2014.

27. [http://www.inegi.org.mx] INEGI. 2014. Registros administrativos: defunciones por homicidio. . Accessed on 15(January 2014).

28. Kay, C. 2008. Reflections on Latin American rural studies in the neoliberal globalization period: a new rurality? Development and Change 39: 915–943. doi:10.1111/j.1467-7660.2008.00518.x

29. King, R, MA Adler, and M Grieves. 2012. Cooperatives as sustainable livelihood strategies in rural Mexico. Bulletin of Latin American Research 32: 166–177. doi:10.1111/j.1470-9856.2012.00796.x

30. Krätli, S, and N Schareika. 2010. Living off uncertainty: the intelligent animal production of dryland pastoralists. European Journal of Development Research 22: 605–622. doi:10.1057/ejdr.2010.41

31. Kristjanson, P, A Krishna, M Radeny, J Kuan, G Quilca, A Sanchez-Urrelo, and C Leon-Velarde. 2007. Poverty dynamics and the role of livestock in the Peruvian Andes. Agricultural Systems 94: 294–308. doi:10.1016/j.agsy.2006.09.009

32. Long, N. 1998. Cambio rural, neoliberalismo y mercantilización. In Las Disputas por el México rural: Actores y campesinos sociales, ed. Sergio Z, Vries P, 45–71. El Colegio de Michoacán AC, Zamora.

33. Mora Ledesma, MI. 2011. Vámonos con todo y chivas. Revista del Colegio de San Luis 1: 49–68.

34. Morand-Fehr, P. 2005. Recent developments in goat nutrition and application: A review. Small Ruminant Research 60: 25–43. doi:10.1016/j.smallrumres.2005.06.004

35. Oba, G. 2012. Harnessing pastoralists' indigenous knowledge for rangeland management: three African case studies. Pastoralism: Research, Policy and Practice 2: 1–25. doi:10.1186/2041-7136-2-1

36. Ochoa, CE. 2013. Feast and famine: The growth of corporate wealth and food insecurity in neoliberal Mexico. Food Sovereignty: A Critical Dialogue. International Conference Yale University,

37. Oseguera Montiel, D, K Frakena, HMJ Udo, NM Keilbach Baer, and AJ van der Zijpp. 2013. Prevalence and risk factors for brucellosis in goats in areas of Mexico with and without brucellosis control campaign. Tropical Animal Health and Production 45: 1383–1389. doi:10.1007/s11250-013-0375-6

38. Peacock, C. 2005. Goats–a pathway out of poverty. Small Ruminant Research 60: 179–186. doi:10.1016/j.smallrumres.2005.06.011

39. Pereyra, G. 2012. México: violencia criminal y "guerra contra el narcotráfico". Revista Mexicana de Sociología 74: 429–460.

40. [http://www.profeco.gob.mx/precios/canasta/default.aspx] PROFECO. 2012. Procuraduría federal del consumidor. Accessed 28 August 2012.

41. 2012. R: A Language and Environment for Statistical Computing. R Foundation for Statistical Computing, Vienna, Austria.

42. Rabell, CA. 1986. Los diezmos de San Luis de la Paz. Instituto de Investigaciones Sociales, Universidad Nacional Autónoma de México, México, D.F..

43. [http://www.ran.gob.mx/ran/index.php/nuestros-sistemas/padron-historico-phina] RAN. 2010. (RegistroAgrarioNacional) Padrón e historial de núcleosagrarios Accessed 14 December 2011.

44. Rodríguez Wallenius, CA. 2011. Dominio del terreno campesino por mineras canadienses 2011. Newspaper reportage, La Jornada del Campo.

45. Schmidt, RH, and WC Gruben. 1992. Ejido Reform and the NAFTA. Federal Reserve Bank of San Francisco: Economic Letter 92. [http://www.frbsf.org/publications/economics/letter/1992/el92-34.pdf] http://www.frbsf.org/publications/economics/letter/1992/el92-34.pdf

46. Scoones, I. 2009. Livelihoods perspectives and rural development. Journal of Peasant Studies 36: 171–196. doi:10.1080/03066150902820503

47. Sinn, R, J Ketzis, and T Chen. 1999. The role of woman in the sheep and goat sector. Small Ruminant Research 34: 259–269. doi:10.1016/S0921-4488(99)00078-4

48. Toledo, V. 1992. Utopía y naturaleza. El nuevo movimiento ecológico de los campesinos e indígenas de América Latina. Nueva Sociedad 122: 72–85. 1992

49. van der Ploeg, JD. 2010. Nuevos campesinos: Campesinos e imperios alimentarios. Icaria Editorial, Barcelona.

50. Wickham, H. 2009. ggplot2: elegant graphics for data analysis. Springer, New York.

51. Zendejas Romero, JS. 2003. Política local y formación del estado: procesos históricos de formación de espacios y sujetos sociales en un municipio rural Mexicano, 1914-1998. Wageningen University, Wageningen.

15

Insect Pests of Sweetpotato in Uganda: Farmers' Perceptions of their Importance and Control Practices

Joshua Sikhu Okonya[1], Robert OM Mwanga[2], Katja Syndikus[3], and Jürgen Kroschel[0]

[1]Global Program of Integrated Crop and Systems Research, International Potato Center (CIP), Plot 47, Ntinda II Road, Naguru, Kampala, Uganda

[2]International Potato Center (CIP), Plot 47, Ntinda II Road, Naguru, Kampala, Uganda

[3]IMOswiss AG, Weststrasse 51, CH-8570 Weinfelden, Switzerland

[4]Global Program of Integrated Crop and Systems Research, International Potato Center (CIP), Avenida La Molina 1895, Lima, 12, Peru

ABSTRACT

Insect pests are among the most important constraints limiting sweetpotato (*Ipomoea batatas*) production in Africa. However, there is inadequate information about farmers' knowledge, perceptions and practices in the management of key insect pests. This has hindered development of effective pest management approaches for smallholder farmers. A standard questionnaire was used to interview individual sweetpotato farmers (n = 192) about their perception and management practices regarding insect pests in six major sweetpotato producing districts of Uganda. The majority (93%) of farmers perceived insect pests to be a very serious problem. With the exception of Masindi and Wakiso districts where the sweetpotato butterfly (*Acraea acerata*) was the number one constraint, sweetpotato weevils (*Cylas puncticollis* and *C. brunneus*) were ranked as the most important insect pests. Insecticide use in sweetpotato fields was very low being highest (28–38% of households) in districts where *A. acerata* infestation is the biggest problem. On average, 65% and 87% of the farmers took no action to control *A. acerata* and *Cylas* spp., respectively. Farmers were more conversant with the presence of and damage by *A. acerata* than of *Cylas* spp. as they thought that *Cylas* spp. root damage was brought about by a prolonged dry season. Different levels of field resistance (ability of a variety to tolerate damage) of sweetpotato landraces to *A. acerata* (eight landraces) and *Cylas* spp. (six landraces) were reported by farmers in all the six districts. This perceived level of resistance to insect damage by landraces needs to be investigated. To improve farmers' capabilities for sweetpotato insect pest management, it is crucial to train them in the basic knowledge of insect pest biology and control.

INTRODUCTION

Sweetpotato (*Ipomoea batatus* L. Lam.) is the world's sixth most important food crop consumed after rice (*Oryza sativa L.*), wheat (*Triticum aestivum* L.), potato (*Solanum tuberosum* L.), maize (*Zea mays* L.), and cassava (*Manihot esculenta* Crantz) (CIP 2010). It is also the third most important root crop grown in eastern Africa after cassava and potato (FAO 2011). Sweetpotato is both a staple and a food security crop in eastern and southern Africa, and is mainly grown by smallholder women farmers (Mutuura et al. 1992; Bashaasha et al. 1995; Andrade et al. 2009). Sweetpotato is also grown for its vines as planting material; leaves are often eaten as a vegetable while shoots and roots are used as animal feed in many countries. In Uganda and western Kenya, the sale of fresh sweetpotato roots, vines and processed foods in both local and urban markets is becoming increasingly popular thus contributing to household cash income (Abidin 2004; Kaguongo et al. 2012). Orange-fleshed sweetpotato is also a rich source of beta-carotene, a precursor of bio-available vitamin A, and has potential of combating Vitamin A deficiency among rural resource-constrained farmers in many developing countries (Jalal et al. 1998; Jaarsveld et al. 2005; Low et al.2007; Mwanga et al. 2003a; Burri 2011).

Sweetpotato production in Uganda declined from 2.84 million tons in 2010 to 2.55 million tons in 2011 (FAO 2011). Uganda also dropped from being the second to being the fourth largest producer of sweetpotato in the world after China, Tanzania and Nigeria (FAO 2011). This reduction in sweetpotato production could be due to many biotic stresses including insect pests and diseases (viral and fungal) that have led to the decline in vine and root quality and root yields. Research conducted in most southern and eastern Africa has shown that insects are among the most economically important pests of sweetpotato (Smit 1997). The most serious and commonly reported insect pest species in Uganda (Smit 1997; Hakiza et al. 2000; Abidin 2004; Ebregt et al.2005), Kenya (Smit 1997; Nderitu et al. 2009), Ethiopia (Azerefegne et al. 2001), Rwanda (Smit et al. 1997) and Nigeria (Girma 1994; Tewe et al. 2003) are caterpillars of the sweetpotato butterfly,*Acraea acerata* Hew. (Lepidoptera: Nymphalidae), the African sweetpotato weevils, *Cylas brunneus*F. and *C. puncticollis* Boheman (Coleoptera: Brentidae), the clearwing moth *Synanthedon* spp. (Lepidoptera: Sesiidae), the sweetpotato hornworm, *Agrius convolvuli* L. (Lepidoptera: Sphingidae) and vectors of the sweetpotato virus diseases, such as the whitefly *Bemisia tabaci* Gennadius (Hemiptera: Aleyrodidae) (Fuglie 2007). *Acraea acerata* and *Cylas* spp. occur in 15 and 23 African countries, respectively (CAB International 2005).

The two African *Cylas* species often occur together in fields and cause huge yield losses of up to 100% (Girma 1994; Smit 1997; Chalfant et al. 1990). Root and stem damage by *Cylas* species is of great economic importance as it leads to a reduction in root yield, and root and vine quality. Due to egg laying and extensive larval tunneling by *Cylas* species, the plant produces bitter defense terpenes and phenolic compounds that make roots unsuitable for both human and animal consumption (Chalfant et al. 1990; Ames et al. 1996).

The *A. acerata* larvae feed on leaves of the sweetpotato plant and heavy infestations can lead to complete plant defoliation. Defoliation of young plants by *A. acerata* larvae can sometimes result in failure of the crop to re-establish and hence reduced yields. As a result, farmers continue to record huge losses due to these pests. Since sweetpotato is an important food security crop, its low production has a bearing impact on national food security. Understanding sweetpotato production constraints, particularly insect pests and farmers' methods of managing *A. acerata* and *Cylas*species, could be useful in designing an effective integrated pest management (IPM) strategy. If sweetpotato IPM research is to be useful and sustainable, it is important to understand farmers' perceptions and their management methods.

Assessing farmers' perceptions of crop production constraints has been used as a tool for documenting pest status and designing pest management options suitable for a particular community (Smit and Matengo 1995; Obopile et al. 2008). Soleri et al. (2000) emphasized the need for integrating methods from both biological and social sciences to understand farmers' selection criteria of crop varieties, since objectives of farmers while selecting a particular maize variety, differed from what research scientists had normally assumed. In a similar way, it would be equally important for crop entomologists to integrate farmers' perspectives of insect pest management in the development of any intervention measure for local use. Bonhof et al. (2001) used farmer participatory rural appraisals for maize farmers at the Kenyan coast to understand the pest status and control strategies for the maize stemborer (*Chilo partellus* Swinhoe). In central Zambia, Mukanga et al. (2011) solicited farmers' views on various management practices of maize ear rots caused by several fungal pathogens before resistant varieties to ear rots could be developed. Lebesa et al. (2012) and Midega et al. (2012) undertook household surveys in western Kenya using questionnaires to determine the pest status of herbivorous blister beetles (*Hycleus* spp.) and cotton (*Gossypium hirsutum* L.) pests. Tounou et al. (2013), similarly used questionnaires to interview farmers to determine geographic distribution and importance of stemborers on maize in southern Togo.

The amount of edible sweetpotato roots and vines damaged by insect pest infestations is significant and calls for an effective management strategy. To develop an integrated pest management strategy that is appropriate for resource poor farmers, information is needed on the current distribution, importance and control methods of sweetpotato pests in various agro-ecologies of Uganda. In Uganda, detailed studies on sweetpotato insect pests are limited. Since the farm surveys conducted by Bashaasha et al. (1995) between 1989 and 1992 in nine different agro-ecological zones, no recent quantitative surveys have been undertaken in the country. The pest status and distribution of key insect pests are expected to vary considerably with the current climate variability. There is, therefore, need to find feasible solutions to agricultural production constraints by incorporating farmer views into research for development programs.

The International Potato Center (CIP) through its global program of integrated crop and systems research seeks to develop an effective IPM strategy for key insect pests of sweetpotato according to climatic factors and pest severity. In this regard, a questionnaire was developed to capture the information and to document the present insect pest distribution, farmers' knowledge and coping strategies to control insect pests of sweetpotato. This study was specifically carried out to 1) determine farmers' perceptions of importance of sweetpotato insect pests and their distribution, 2) analyze the major sweetpotato production constraints and farmers' coping strategies for the control of sweetpotato insect pests in various agro-ecologies of Uganda, 3) identify local sweetpotato landraces that have some field resistance to damage by *A. acerata* and *Cylas* spp.

MATERIALS AND METHODS

Study Area

The study was conducted between August and October 2011 in six districts of Uganda. Major sweetpotato growing districts were purposively selected from six different agro-ecological zones to obtain a wide range of household perceptions. These were Gulu (2°46′48″N, 32°18′0″E), Kabale (1° 15′ 0″ S, 29° 59′ 24″ E), Kasese (0° 11′ 0″ N, 30° 5′ 0″ E), Masindi (1° 41′ 1″ N, 32° 43′ 20″ E), Wakiso (0°23′53″N, 32°28′41″E) and Soroti (1° 42′ 54″ N, 33° 36′ 40″ E) (Figure 1). In each district, two sub-counties were purposively selected based on intensity of sweetpotato farming as perceived by the district agriculture office for this observational study.

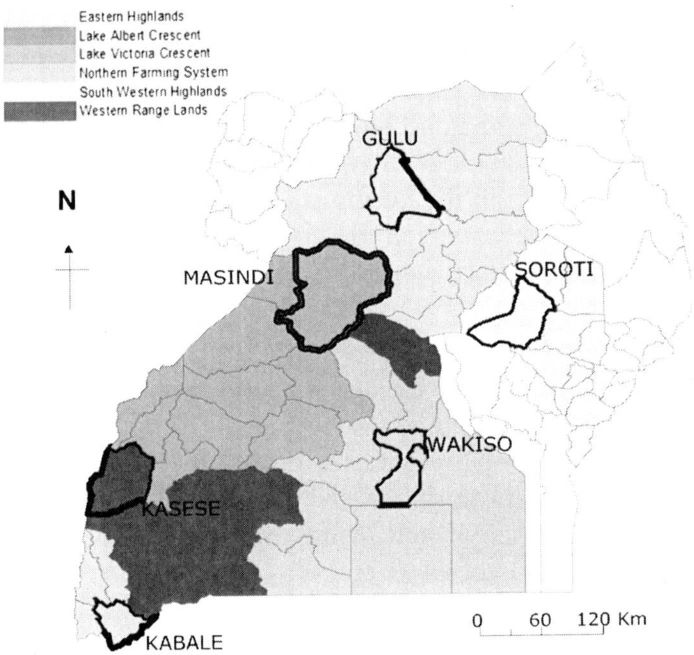

Figure 1: Map of Uganda showing the six study districts with their corresponding agro-ecological zones; Gulu in the northern farming system, Masindi in the Lake Albert Crescent, Soroti in the Eastern Highlands, Wakiso in the Lake Victoria crescent, Kasese in the western range highlands and Kabale in the south western highlands.

The questionnaire was pre-tested with five sweetpotato farmers in Luweero district, not included in the sample, three weeks before the study. After the pilot test, changes were made in the expression of some questions to be asked. The survey team consisted of two entomologists, one research assistant and twelve agricultural extension workers.

Farmers who had grown sweetpotato in 2011 and preferably still had it in their fields were randomly selected at regular intervals of 5–10 km distance between sites or until a sweetpotato growing household was found on motorable roads within the subcounties. A total of 32, both female and male farmers, were interviewed face-to-face in their homesteads in each district using a standard questionnaire – partly structured and partly open. All interviews were conducted in the local language of that district with the help of an enumerator. Data was collected on: sweetpotato cropping systems, constraints in sweetpotato production, sweetpotato insect pests and, their control measures and field resistance of sweetpotato varieties to the three major insect pests (*A. acerata, C. brunneus* and *C. puncticollis*).

Constraints to sweetpotato production were identified by asking farmers to mention all the constraints they had experienced, and ranking them in order of importance of each constraint (for instance most destructive, occurs frequently) in the last cropping season. Farmers were shown colored photographs and/or vials of insects (larval and adult stages) in alcohol to ensure they made the correct identification of the pest. Farmers' compared the perceived field resistance (level of damage) of their presently cultivated local varieties to damage by the most important insect pests.

Quantitative data were subjected to analysis of variance (ANOVA) using General Linear Model (GLM) in SAS software (SAS Institute Inc 2008). Means were compared using Fisher's LSD Test. Descriptive statistics (percentages and mean values) were the main statistical tools employed to analyze qualitative data.

RESULTS AND DISCUSSION

Farm Household Characteristics and Some Aspects of Sweetpotato Production

Sweetpotato production in the six districts of Uganda is a female dominated activity with males representing only 39% of 192 respondents. In Soroti district, however, male respondents outnumbered female respondents by 25%; this is probably because sweetpotato is more of a commercial than food security crop in this district. Men have been reported in many African countries to dominate cash crop production (World Bank et al. 2009). This domain change of sweetpotato, from being a female controlled crop to a male crop in Soroti due to commercialization, has been reported for other crops like coffee (*Coffea* spp.) in Uganda (Kasente et al. 2002) and cocoa (*Theobroma cacao* L.) in Ghana (Doss 2001). In Uganda, 51% of the population is female (UNHS 2010) and women are mostly responsible for household food crop production; in particular the sweetpotato crop which is referred to as a "female" crop (Bashaasha et al. 1995). Consequently, efforts geared towards promoting and disseminating new sweetpotato technologies should mainly target women.

Most of the sweetpotato farmers were in their active age (mean of 40 to 46 years) which is a good sign for food security in Uganda (Table 1). There is a huge potential to increase production if more investments are made into developing pest and disease resistant high yielding sweetpotato varieties as part of developing integrated pest management strategies. Mean household size across districts was 7.5 members; this is higher than the national average by 2.5 members (UNHS 2010). On average, Gulu district had the biggest households with 10 members, while Kabale district had the smallest household size of six members (Table 1).

Table 1: Household characteristics of interviewed sweetpotato farmers in the six districts of Uganda and their sweetpotato production practices/techniques, August-October 2011

Household characteristics	Gulu	Kabale	Kasese	Masindi	Soroti	Wakiso	p-value (N/A = not applicable)
Female respondents (%)	53.1	62.5	59.4	75.0	37.5	78.1	N/A
Female headed households (%)	28.1	28.1	21.9	28.1	15.6	43.8	N/A
No formal education (%)	28.1	25.0	18.8	9.4	6.3	20.0	N/A
≤7 years of formal education (%)	40.6	65.6	56.3	46.9	71.9	60.0	N/A
>7 years of formal education (%)	31.3	9.4	25.0	43.8	21.9	20.0	N/A
Mean elevation (m a.s.l.)	1091^{cd}	2090^{a}	1006^{e}	1124^{c}	1079^{d}	1191^{b}	<0.0001
Mean age of respondent (years)	43.7 ± 2.6^{a}	40.0 ± 2.0^{a}	41.2 ± 2.4^{a}	44.9 ± 2.4^{a}	44.1 ± 2.0^{a}	45.8 ± 2.4^{a}	0.7024
Mean household size (persons)	10.2 ± 1.3^{a}	5.9 ± 0.5^{c}	6.6 ± 0.5^{bc}	6.5 ± 0.6^{c}	8.6 ± 0.7^{ab}	7.5 ± 0.8^{bc}	0.0002

Mean rotation duration (months)	20.4 ± 2.2^{ab}	11.0 ± 1.7^{c}	6.6 ± 0.8^{c}	17.2 ± 1.4^{b}	23.3 ± 2.1^{a}	9.1 ± 1.1^{c}	<0.0001
Mean years of growing sweetpotato	25.1 ± 2.7^{ab}	19.6 ± 1.9^{abc}	16.8 ± 2.0^{c}	26.3 ± 2.3^{a}	23.8 ± 1.9^{ab}	19.7 ± 2.4^{bc}	0.0179
Mean sweetpotato acreage in 2011 (ha)	0.3 ± 0.1^{ab}	0.2 ± 0.0^{b}	0.2 ± 0.0^{b}	0.2 ± 0.0^{b}	0.3 ± 0.0^{ab}	0.4 ± 0.1^{a}	0.1393
Arable land devoted to sweetpotato (%)	12.3 ± 3.1^{b}	30.5 ± 5.0^{a}	17.9 ± 3.8^{b}	14.1 ± 2.0^{b}	20.0 ± 3.5^{b}	31.1 ± 5.1^{a}	<0.0001
Mean total land holding (ha)	100.2 ± 82.3^{a}	2.9 ± 1.7^{a}	2.4 ± 0.5^{a}	6.6 ± 2.8^{a}	2.4 ± 0.3^{a}	1.3 ± 0.2^{a}	0.2625
Total land cropped (%)	59.5 ± 6.3^{c}	85.8 ± 5.6^{ab}	85.4 ± 3.9^{ab}	63.6 ± 5.9^{c}	77.8 ± 5.6^{b}	93.9 ± 3.0^{a}	<0.0001

Means followed by the same letter in the same row are not significantly different ($p \geq 0.05$, Fisher's least significant difference). Values are means ± SE.

Okonya *et al. SpringerPlus* 2014 3:303, doi: 10.1186/2193-1801-3-303

Mean elevation of the surveyed homesteads was highest in Kabale district (2090 m a.s.l.) and lowest in Kasese district (1005 m a.s.l.). Sweetpotato had been grown for more than 16 years in all the six districts. The average period of growing sweetpotato was longest (26 years) for Masindi followed by Gulu district (25 years). The average sweetpotato acreage in 2011 was highest in Soroti district (0.4 ha) and lowest in Kabale, Kasese and Masindi districts (0.2 ha). Sweetpotato is a commercial crop in Soroti district and this could explain the allocation of relatively larger portions of arable land to sweetpotato production.

Constraints to sweet potato production

Identification of factors limiting production and provision of environmentally-friendly options for integrated crop management is inevitable if sweetpotato production among smallholder farmers in Uganda is to be increased. In the districts visited, setbacks to sweetpotato production were many, however, insect pests attacking roots and leaves were the most important (Table 2). Poor yields of local varieties and degraded soils came second in importance while rodents or rats were ranked the third most important constraint. Other constraints that were unique to a district included delayed rainfall and weeds in Wakiso, shortage of land and small size of roots in Kabale, high cost of labor, floods, millipedes (Diplopoda) and poor roads in Kasese, extreme rainfall, high transport cost, lack of capital to plant large fields and high labor demands for the sweetpotato crop in Gulu districts.

Table 2: Top five most important constraints to sweetpotato production as ranked by farmers in the study districts of Uganda, August-October 2011

Constraints to sweetpotato production	Rank for each constraint (% households)*					
	1st	**2nd**	**3rd**	**4th**	**5th**	**Mean**
No constraint mentioned/ experienced	0.5	8.3	26.6	55.7	81.8	34.6
Insect pests	33.9	42.2	33.3	18.8	5.7	26.8

Poor yields of local varieties/soils	12.0	10.4	6.3	1.6	1.6	6.4
Rats and rodents	8.3	8.9	6.3	4.2	1.0	5.7
Drought/prolonged dry seasons	10.9	3.1	4.2	2.6	0.5	4.3
High cost of labor/ labor intensive/ shortage of labor/ weeds	6.3	2.1	2.6	1.6	1.6	2.8
Lack of market	3.6	5.2	2.1	2.1	0.5	2.7
Viral diseases	2.6	3.1	3.1	2.6	1.0	2.5
Shortage of planting material	5.2	1.6	3.1	1.0	0.0	2.2
Wild game (elephants, hippos, pigs)	4.2	2.6	1.6	1.0	1.0	2.1
Others	1.5	2.6	2.1	3.1	1.0	2.1
Millipedes	1.6	1.6	3.6	0.5	2.6	2.0
Land shortage	2.1	2.1	2.6	0.5	0.5	1.6
Floods/excess rainfall/ storms	1.0	1.6	1.0	1.0	0.5	1.0
Lack of money to hire labor/plant large fields/build drying places	2.1	1.6	0.0	0.5	0.0	0.8
Domestic animals (cattle, goats)	1.0	1.0	0.0	1.0	0.5	0.7
Changed onset and cessation of rainfall in the seasons	1.6	0.5	0.5	0.5	0.0	0.6
Alternaria blight disease	1.0	0.5	0.0	1.0	0.0	0.5
Root rots	0.0	1.0	1.0	0.5	0.0	0.5

Some columns do not add up to 100% due to rounding off.

Okonya *et al.SpringerPlus* 2014 3:303, doi: 10.1186/2193-1801-3-303

Sweetpotato Insect Pests

Among the sweetpotato insect pests reported by farmers, *Cylas* spp. were ranked number one by 87% of the households followed by caterpillars of *A. acerata* (60.7%) (Table 3). Caterpillars of *A. acerata* in Masindi and Wakiso districts were perceived to be more damaging to sweetpotato than the *Cylas* spp. Insect pests that appeared to be local included *Agrius convolvuli* in Gulu and Soroti districts, and the sweetpotato armyworm (*Spodoptera* spp.) in Soroti district. *A. convolvuli*occasionally appeared in both districts as in 2009 and 2010 where it caused serious sweetpotato damage leading to food insecurity (IPC 2010). *Spodoptera* spp. had occurred in Soroti in the past three consecutive years (2007 to 2010). Small black ants ("*Munyeera*" in Luganda) were mentioned to construct nests inside the stem base of sweetpotato plants leading to vine damage. Ants are not pests *per se* but rather predators of *Cylas* spp. larvae feeding inside the stem base.

Table 3: Major insect pests experienced by farmers in sweetpotato (% households) and rank

Insect pest reported	% households*						
	Gulu	Kabale	Kasese	**Masindi**	**Soroti**	**Wakiso**	**Overall mean**
Sweetpotato weevils (*Cylas*spp.)	97 (1)	56 (1)	84 (1)	100 (2)	94 (1)	91 (1)	87.0
Sweetpotato butterfly (*Acraea acerata*)	28 (2)	50 (2)	88 (2)	88 (1)	13 (2)	97 (2)	60.7
Sweetpotato hornworm (*Agrius convolvuli*)	33	0	0	0	25	0	9.7
Armyworm (*Spodoptera*spp.)	0	0	0	0	34	0	5.7
Others (ants, whiteflies)	0	0	9	3	3	0	2.5

*Percentage values add to more than 100 due to multiple responses. Rank (in parentheses) for the top two most important insect pests.

Okonya *et al. SpringerPlus* 2014 **3**:303, doi: 10.1186/2193-1801-3-303

Our results are in accordance with those reported by Girma and Belehu (1994) in Ethiopia where insect pests and specifically *A. acerata* and *Cylas* spp. are most important. Farmers reported that *A. acerata* was important only in some years with the last distinct outbreak in Soroti being in February 2010. Farmers in Masindi, however, said that *A. acerata* started to be a problem in 1993 until now.

It is important to note that farmers tend to have a high perception of damage caused by insect pests, hence, rank them highly (Smit 1997). This must have been the case in the current study since results from a subsequent field survey of farmers' sweetpotato fields in Kabale, Buliisa and Masindi districts reported lower insect pest infestation rates and densities (Okonya and Kroschel2013a, b). *B. tabaci* is an important insect vector of the sweetpotato virus disease component (*Sweetpotato chlorotic stunt virus*) in Uganda. However, most farmers did not consider *B. tabaci* an important insect pest because they do not have the knowledge to relate the presence of the insect (vector) to virus symptoms or crop damage and yield loss. Such knowledge gaps call for better training on IPM through the national extension services.

There was a general belief by farmers that once sweetpotato is attacked by *A. acerata,* root yield will be poor. The African locust (*Locusta migratoria migratorioides* (R. & F.) appeared in Soroti district in 2008/2009 cropping season and completely defoliated sweetpotato plants.

Farmers Management Practices of Sweetpotato Pests

The main methods used by farmers in managing sweet potato·insect pests on their farms included chemical insecticide application, ash application, hand-picking or a combination of two or more physical and cultural strategies (Table 4). Control strategies for caterpillars of *A. acerata* included use of chemical insecticides of mainly permethrin, dimethoate and cypermethrin by 24% of the households; application of wood ash, hand-picking of caterpillars and a combination of two or three of the above mentioned methods. Early harvesting was the most common method used to limit the damage caused by *Cylas* spp., however, mulching, re-hilling to cover soil cracks, crop rotation and insecticide application were being used but on a very low scale. On average, 65% and 87% of the sweetpotato farmers did not control *A. acerata* and *Cylas* spp., respectively.

Table 4: Control methods for the two (A. *acerata* and *Cylas* spp.) most important insect pests of sweetpotato (% households)

Control strategy for *A. Acerata* and *Cylas* spp.	% households						Overall mean
	Gulu	Kabale	Kasese	Masindi	Soroti	Wakiso	
a) for *A. acerata*							
Chemical insecticides	2	20	14	50	0	56	24
Ash application	0	0	7	8	0	0	3
Hand-picking	0	0	0	8	0	6	2
Chemical insecticides and hand-picking	0	0	0	8	0	0	1
Chemical insecticides and ash application	0	0	0	0	0	13	2
Hand-picking and ash application	0	0	0	8	0	0	1
Chemical insecticides, hand-picking and ash application	0	0	0	0	0	6	1
None	98	80	79	17	100	19	65
b) For *Cylas*spp.							
Chemical insecticides	0	0	0	6	0	8	2
Chemical insecticides and re-hilling	7	0	7	6	0	0	3
Crop rotation	7	8	0	0	0	0	3

Early harvesting	0	0	7	0	7	8	4
Mulching and re-hilling	0	0	0	0	0	8	1
None	87	92	87	88	93	77	87

Okonya *et al.*

Okonya *et al.SpringerPlus* 2014 **3**:303, doi: 10.1186/2193-1801-3-303

Use of chemical insecticides was relatively low; being highest in Wakiso district followed by Masindi district but absent in Soroti district (Table. 4). Insecticide application was highest in districts which ranked *A. acerata* as the main insect pest damaging sweetpotato. It was evident that some farmers implemented more than one control strategy to reduce field infestation by insect pests. Aiming at improving the effectiveness of available control methods is therefore desired. The low use of insecticides in sweetpotato could partly be due to the high cost of insecticides which subsistence farmers cannot afford but also lack of knowledge about pest biology of especially *Cylas* spp. Many farmers did not know how to control *Cylas* spp. However, farmers who applied insecticides to control *A. acerata* observed reduced damage by *Cylas* spp. as well. This therefore encouraged farmers to apply insecticides two months after planting even in the absence of *A. acerata*.

Due to the severity of pest infestation during outbreaks, farmers usually received insecticides from agriculture extension workers or local authorities to spray against *A. acerata* in Kabale district and*A. convolvuli* in Gulu and Soroti districts. Use of chemical insecticides against *A. acerata* could be because farmers took the pest seriously, pest and pest damage were more visible or pesticides are effective. It should be noted, however, that use of chemical insecticides is not a permanent solution as it can be disastrous to human health due to poor handling, elimination of natural enemies for the pest and is out of reach for most of the resource constrained poor farmers (Croft and Brown1975).

Major Sweetpotato Varieties Grown and Farmers Perceptions of Their Field Resistance to Insect Pests

Sweetpotato variety mixtures were a common practice by at least 89% of interviewed farm households. With the exception of households in Kabale and Soroti districts, farmers in the remaining districts sometimes planted vines affected by insect pests due to lack of planting materials, especially at the end of a prolonged dry season. The most popular grown varieties were Larila in Gulu, Rwabafuluki in Kabale, Rwatoro in Kasese, Dimbuka in Masindi, Kampala and Araka in Soroti and NASPOT 1 in Wakiso district (Table 5). Farmers identified eight landraces (Mukono, Kigabali/Magabali, Red mamba, Kiryenamwami, Dimbuka, Boy, Setyabule and Mbale), which have shown some form of field resistance to vine damage by *A. acerata* and six landraces (Ochol/Ocuc, Rwatoro, Muhamoud, Dimbuka, Kyebandula and Opaku) with some form of resistance to field root damage by *Cylas* spp. Information on resistance levels of landraces needs to be taken with caution since no variety has been reported to be resistant in laboratory no choice experiments to *Cylas* spp. for instance (Mwanga et al. 2003a, b). This notwithstanding, these reports of resistance of landraces to insects need to be investigated further as they may provide potential sources of resistance to these two most economically important pests of sweetpotato. Various authors have found differences in *Cylas* spp. damage among cultivars (Mwanga et al. 2003b; Stathers et al.2003). Factors such as quantity of root latex, depth of rooting and amount of foliage have been reported to contribute to reduced *Cylas* spp. field damage (Mwanga et al. 2001; Stathers et al.2003). Anyanga et al. (2013) found that chemical compounds in the root latex were responsible for the host plant resistance to *Cylas* spp. damage of "New Kawogo" sweetpotato variety. This variety (New Kawogo) was also mentioned by farmers in Masindi and Wakiso districts in this study to be resistant to *Cylas* spp. damage.

Table 5: Perceived field resistance of major local sweetpotato varieties or landraces to*A. acerata* and *Cylas* spp. in 2011

District	Local name of Sweetpotato variety/ landrace	% Households growing the variety	Resistence level to*Acraea acerata* damage(%responses)			Resistence level to *Cylas*spp. damage (% responses)		
			Low	Moderate	High	Low	Moderate	High
GULU	Lalira	22	0	15	9	0	13	11
	Alero	12	0	6	3	0	7	4
	Ochol/Ocuc	12	0	6	3	1	7	1
	Adoch	11	0	6	15	0	2	9
	Mukiga	3	0	3	0	0	2	2
	Others	40	0	18	18	0	18	21
KABALE	Rwabafuluki, Kandazi/ Mulungi	22	0	4	25	0	7	17
	Mukazi	17	0	0	11	0	17	14
	Mukono	8	4	0	11	0	7	3
	Kidodo	6	0	4	11	0	3	0
	Kigabali/ Magabali	4	4	0	4	0	7	0
	Others	43	0	0	25	0	17	7

KASESE	Rwatoro	13	0	2	9	2	2	6
	Red mamba	10	2	4	4	0	4	0
	Rosemary	10	0	4	7	0	6	6
	Kiryenamwami	8	2	2	7	0	4	4
	Muhamoud	6	0	4	7	2	2	6
	Bitambi	6	0	2	2	0	2	2
	Kyebandula	6	0	2	2	0	4	2
	Others	41	11	0	26	4	21	17
MASINDI	Dimbuka	22	1	7	18	1	16	10
	Nakato/ Nyakato	14	0	6	12	0	14	5
	Kahogo/New Kawogo	9	0	4	9	0	5	4
	Kyebandula	5	0	1	4	1	1	0
	Suwedi	3	0	3	1	0	3	3
	Kabakumba	3	0	0	3	0	0	3
	Others	44	0	4	25	0	18	16

SOROTI	Kampala	18	0	0	11	0	6	12
	Araka	18	0	5	16	0	8	12
	Ateseke	7	0	5	11	0	6	4
	Opaku	5	0	0	0	1	4	1
	Letesi/Latesi	4	0	0	0	0	3	1
	Mwambi	4	0	5	11	0	3	1
	Boy	4	5	0	0	0	0	4
	Others	40	0	16	16	1	16	17
WAKISO	Naspot 1	31	0	16	23	0	18	35
	Dimbuka	23	2	5	16	3	3	15
	Setyabule	13	2	5	7	0	3	3
	Mbale	10	2	0	7	0	3	5
	New Kawogo	6	0	2	2	0	5	5
	Others	17	0	7	5	0	5	0

Okonya et al.

Okonya *et al.SpringerPlus* 2014 **3**:303 doi: 10.1186/2193-1801-3-303

The study shows that sweetpotato varieties susceptible to insect pests are cultivated by farmers across the six districts. This is due to the fact that high resistance to either *A. acerata* or *Cylas* spp. has not been found in the world sweetpotato germplasm collection (Mwanga et al. 2009), which could have been used to develop and release resistant varieties in Uganda. Evaluations for resistance against *C. puncticollis* of transgenic sweetpotato expressing Cry3Ca1, Cry7Aa1 and ET33-34 proteins from *Bacillus thuringiensis* (*Bt*) at the National Crop Resources Research Institute, Namulonge, Uganda are underway (Rukarwa et al. 2013).

CONCLUSIONS AND OUTLOOK

This study has provided insight into sweetpotato production in the main production districts of Uganda. Insect pests are a major production constraint in all the districts surveyed. The majority of farmers perceive insect pests to be a very serious problem. With the exception of Masindi and Wakiso districts where *A. acerata* is the number one constraint, sweetpotato weevils (*C. puncticollis*and *C. brunneus*) are ranked as the most important insect pests. However, many sweetpotato farmers take no measures to control *Cylas* spp. but invest in the use of chemical insecticides to control *A. acerata*, which has a high priority in Masindi and Wakiso districts. Promoting the use of cultural control methods such as mulching, re-hilling to cover soil cracks, use of clean planting material, crop rotation, taking time period and crops planted before into account, have potential to reduce damage by *Cylas* spp. Further, it would be important to invest in research to develop additional control measures. Biological control using the entomopathogen *Beauveria bassiana* has shown to reduce the *Cylas formicarius* Fabricius damage in Cuba alone or in combination with sexual pheromones (Lagnaoui et al. 2000) Farmers lack knowledge on *Cylas* spp. biology, damage and control. Hence, training about the biology and ecology of this important pest can help farmers to make informed crop management decisions. Baseline information gained from this study will assist the international agricultural research system (NARS), and non-governmental organizations (NGO's) in designing IPM strategies that are based on the needs of smallholder farmers and their sweetpotato production systems. This information will also be useful in setting research priorities.

AUTHORS' CONTRIBUTIONS

JSO designed the questionnaire, collected and analyzed the data, and drafted the manuscript. KS participated in designing the questionnaire, data collection and edited the manuscript; ROMM edited the manuscript and provided overall administrative support for the study. JK conceived and designed the study, revised and polished the manuscript, and supervised the whole study; all authors read and approved the final manuscript.

ACKNOWLEDGEMENT

The authors are thankful to the farmers who participated in the survey, the enumerators for translating the questions into local languages, the district agriculture officers and extension workers in the surveyed districts for assisting in site selection. The assistance of Ssempa Zaamu (CIP-Uganda) during data collection is highly appreciated. This work was supported by a grant from the German Federal Ministry for Economic Cooperation and Development (BMZ), which is highly acknowledged.

REFERENCES

1. Abidin PE (2004) Sweetpotato breeding for northeastern Uganda: farmer varieties, farmer-participatory selection, and stability of performance. Dissertation, Wageningen University
2. Ames T, Smit NEJM, Braun AR, O'Sullivan JN, Skoglund LG (1996) Sweetpotato: major pests, diseases and nutritional disorders. International Potato Center (CIP), Lima, Peru.
3. Andrade M, Barker I, Cole D, Dapaah H, Elliott H, Fuentes S, Grüneberg W, Kapinga R, Kroschel J, Labarta R, Lemaga B, Loechl C, Low J, Lynam J, Mwanga R, Ortiz O, Oswald A, Thiele G (2009) Unleashing the potential of sweetpotato in sub-Saharan Africa: current challenges and way forward. International Potato Center (CIP), Lima, Peru.

4. Anyanga MO, Muyinza H, Talwana H, Hall DR, Farman DI, Ssemakula GN, Mwanga ROM, Stevenson PC (2013) Resistance to the weevils *Cylas puncticollis* and *Cylas brunneus* conferred by sweetpotato root surface compounds. J Agric Food Chem 61(34):8141-8147 PubMed Abstract |

5. Azerefegne F, Solbreck C, Ives AR (2001) Environmental forcing and high amplitude fluctuations in the population dynamics of the tropical butterfly *Acraea acerata* (Lepidoptera: Nymphalidae). J Anim Ecol 70:1032-1045

6. Bashaasha B, Mwanga ROM, Ocitti p'Obwoya C, Ewell PT (1995) Sweetpotato in farming and food systems of Uganda. A farm survey report.Sub-Saharan African region, Nairobi, Kenya/ National Agricultural Research Organization, Kampala, Uganda

7. Bonhof MJ, van Huis A, Kiros FG, Dibogo N (2001) Farmers' perceptions of importance, control methods and natural enemies of maize stemborers at the Kenya coast. Insect Sci Appl 21:33-43

8. Burri BJ (2011) Evaluating sweetpotato as an intervention food to prevent vitamin A deficiency. Compr Rev Food Sci Food Saf 10:118-130

9. CAB International (2005) Distribution maps of plant pests. 1st revision A: map no. 279. Cylas puncticollis (Boheman). 1-2

10. Chalfant RB, Jansson RK, Dakshina RS, Schalk JM (1990) Ecology and management of sweet potato insects. Annu Rev Entomol 35:157-180

11. CIP (International Potato Center) (2010) Facts and figures about sweetpotato 2010. http://cipotato.org/resources/publications/fact-sheets-flyer-leaflet/facts-and-figures-about-sweetpotato/*webcite*. Accessed 10 July 2013

12. Croft BA, Brown AWA (1975) Responses of arthropod natural enemies to insecticides. Annu Rev Entomol 20:285-335

13. Doss C (2001) Designing agricultural technology for African women farmers: lessons from 25 years of experience. World Dev 29:2075-2092

14. Ebregt E, Struik PC, Odongo B, Abidin PE (2005) Pest damage in sweetpotato, groundnut and maize in north-eastern Uganda with special reference to damage by millipedes (Diplopoda). NJAS-Wagen J Life Sci 53:49-69.

15. FAO (Food and Agriculture Organization) (2011) Production quantity of primary crops, country statistics for eastern Africa. http://faostat3.fao.org/faostat-gateway/go/to/download/Q/QC/E. Accessed 29 April 2013

16. Fuglie KO (2007) Priorities for sweetpotato research in developing countries: results of a survey. Hortscience 42:1200-1206

17. Girma A (1994) Review of research on sweetpotato pest management in Ethiopia. In: Allard GB, Skoglund LG, Neuenschwander P, Murphy RJ (eds) Proceedings of a regional workshop on root and tuber pest management in east and southern Africa. Mombasa, Kenya, 10–14 August 1992, Nairobi: IIBC. pp 53-56

18. Girma A, Belehu T (1994) Sweetpotato production constraints and research in Ethiopia. In: Allard GB, Skoglund LG, Neuenschwander P, Murphy RJ (eds) Proceedings of a regional workshop on root and tuber pest management in east and southern Africa. Mombasa, Kenya, 10–14 August 1992, Nairobi: IIBC. pp 49-51

19. Hakiza JJ, Turyamureeba G, Kakuhenzire RM, Odongo B, Mwanga ROM (2000) Potato and sweetpotato improvement in Uganda: a historical perspective. Proceedings of the 5th African potato association (APA) conference, 29 May - 2 June 2000, Makerere University, Kampala. 47-58

20. IPC (Integrated food security Phase Classification) (2010) Uganda food security brief. Food security and agricultural livelihoods cluster. Food and Agriculture Organization, Kampala, Uganda.

21. Jaarsveld PJ, Faber M, Tanumihardjo SA, Nestel P, Lombard CJ, Benadé AJS (2005) -carotene–rich orange-fleshed sweet potato improves the vitamin A status of primary school children assessed with the modified-relative-dose–response test1,2,3. Am J Clin Nutr 81:1080-1087 PubMed Abstract|

22. Jalal F, Nesheim MC, Agus Z, Sanjur D, Habicht JP (1998) Serum retinol concentrations in children are affected by food sources of -carotene, fat intake, and anthelmintic drug treatment. Am J Clin Nutr 68:623-629

23. Kaguongo W, Ortmann G, Wale E, Darroch M, Low J (2012) Factors influencing adoption and intensity of adoption of orange flesh sweetpotato varieties: evidence from an extension

intervention in Nyanza and western provinces, Kenya. Afr J Agric Res 7:493-503

24. Kasente D, Lockwood M, Vivian J, Whithead A (2002) Gender and the expansion of non-traditional agricultural exports in Uganda. In: Razavi S (ed) Shifting burdens: gender and agrarian change under neoliberalism, Bloomfield, CT, USA: Kumarian Press Inc. pp 35-65
25. Lagnaoui FC, Alcázar J, Morales F (2000) A sustainable pest management strategy for the sweetpotato weevil in Cuba: a success story. Food and Fertilizer Technology Center, Extension Bulletin 493:1-7
26. Lebesa LN, Khan ZR, Krueger K, Bruce JA, Hassanali A, Pickett JA (2012) Farmers' knowledge and perceptions of blister beetles, *Hycleus* spp. (Coleoptera: Meloidae), as pest herbivores of*Desmodium* legumes in western Kenya. Int J Pest Manag 58:165-174
27. Low JW, Arimond M, Osman N, Cunguara B, Zano F, Tschirley D (2007) A food-based approach introducing orange-fleshed sweetpotatoes increased vitamin A intake and serum retinol concentrations in young children in rural Mozambique. J Nutr 137:1320-1327
28. Midega CAO, Nyang'au IM, Pittchar J, Birkett MA, Pickett JA, Borges M, Khan ZR (2012) Farmers' perceptions of cotton pests and their management in Western Kenya. Crop Prot 42:193-201
29. Mukanga M, Derera J, Tongoona P, Laing MD (2011) Farmers' perceptions and management of maize ear rots and their implications for breeding for resistance. Afr J Agric Res 6:4544-4554
30. Mutuura J, Ewell PT, Abubaker A, Munga T, Ajanga S, Irungu J, Omari F, Maobe S (1992) Sweetpotato in the food systems of Kenya. Results of a socio-economic survey. Proceedings of KARI/CIP technical workshop on collaborative research, Nairobi.
31. Mwanga ROM, Odongo B, Ocitti p'Obwoya C (2001) Release of five sweetpotato cultivars in Uganda. HortSci 36:385-386
32. Mwanga ROM, Niringiye C, Odongo B, Opio AF, Lemaga B, Kyaddondo B, Sserunjogi L, Owori C, Kapinga R (2003) Partnership: an approach to promote orange-fleshed sweetpotato,

experiences from Uganda. In: Kapinga R, Kingamkono R, Msabaha M, Ndunguru J, Lemaga B, Tusiime G (eds) Proceedings of the 13th international society for tropical root crops (ISTRC) symposium. Tropical root and tuber crops: opportunities for poverty alleviation and sustainable livelihoods in the developing world. November 10–14, 2003, Arusha, pp. 770-779

33. Mwanga ROM, Odongo B, Turyamureeba G, Alajo A, Yencho GC, Gibson RW, Smit NEJM, Carey EE (2003) Release of six sweetpotato cultivars. HortSci 3:475-476

34. Mwanga ROM, Odongo B, Niringiye C, Alajo A, Kigozi B, Makumbi R, Lugwana E, Namakula J, Mpembe I, Kapinga R, Lemaga B, Nsumba J, Tumwegamire T, Yencho CG (2009) 'NASPOT 7', 'NASPOT 8', 'NASPOT 9 O', 'NASPOT 10 O', and 'Dimbuka-Bukulula' Sweetpotato. Hortscience 44:828-832

35. Nderitu J, Silai M, Nyamasyo G, Kasina M (2009) Insect species associated with sweetpotatoes (*Ipomoea batatas* (L.) Lam.) in eastern Kenya. Int J Sustain Crop Prod 4:14-18

36. Obopile MD, Munthali C, Matilo B (2008) Farmers' knowledge, perceptions and management of vegetable pests and diseases in Botswana. Crop Prot 27:1220-1224

37. Okonya JS, Kroschel J (2013) Incidence, abundance and damage by the sweet potato butterfly (*Acraea acerata* Hew.) and the African sweet potato weevils (*Cylas* spp.) across an altitude gradient in Kabale district, Uganda. Int J AgriScience 3:814-824

38. Okonya JS, Kroschel J (2013) Pest status of *Acraea acerata* Hew. and *Cylas* spp. in sweetpotato (*Ipomoea batatas* (L.) Lam.) and incidence of natural enemies in the Lake Albert crescent agro-ecological zone of Uganda. Int J Insect Sci 5:41-46

39. Rukarwa RJ, Prentice K, Ormachea M, Kreuze JF, Tovar J, Mukasa SB, Ssemakula G, Mwanga ROM, Ghislain M (2013) Evaluation of bioassays for testing *Bt* sweetpotato events against sweetpotato weevils. Afr Crop Sci J 21:235-244

40. SAS Institute Inc (2008) The SAS system for windows v9.2. SAS Institute Inc, Cary, North Carolina.

41. Smit NEJM (1997) Integrated pest management for sweetpotato in Eastern Africa. Dissertation, Wageningen University

42. Smit NEJM, Matengo LO (1995) Farmers' cultural practices and their effects on pest control in sweetpotato in South Nyanza Kenya. Int J Pest Manage 41:2-7
43. Smit NEJM, Lugojja F, Ogenga L, Morris W (1997) The sweetpotato butterfly (*Acraea acerata*): a review. Int J Pest Manage 43:275-278
44. Soleri D, Smith SE, Cleveland DA (2000) Evaluating the potential for farmer and plant breeder collaboration: a case study of farmer maize selection in Oaxaca, Mexico. Euphytica 116:41-57
45. Stathers TE, Rees D, Kabi S, Mbilinyi L, Smit N, Kiozya H, Jeremiah S, Nyango A, Jeffries D (2003) Sweetpotato infestation by *Cylas* spp. in east Africa. I cultivar differences in field infestation and the role of plant factors. Int J Pest Manage 49:131-140
46. Tewe OO, Ojeniyi FE, Abu OA (2003) Sweetpotato production, utilization, and marketing in Nigeria. International Potato Center (CIP), Lima, Peru.
47. Tounou AK, Agboka K, Agbodzavu KM, Wegbe K (2013) Maize stemborers distribution, their natural enemies and farmers' perception on climate change and stemborers in southern Togo. J Appl Biosci 64:4773-4786
48. UNHS (Uganda National Household Survey) (2010) Abridged report for the socioeconomic module. Uganda Bureau of Statistics (UBOS), Kampala, Uganda.
49. World Bank (2009) Gender in agriculture: sourcebook. World Bank Publications, Washington, DC.

Chapter 6

Conservation and Management of Sacred Groves, Myths and Beliefs of Tribal Communities: A Case Study from North-India

Laxman Singh Kandari[1], Vinod Kumar Bisht[2], Meenakshl Bhardwaj[3], and Ashok Kumar Thakur[4]

[1]School of Natural Resource Management and Environmental Sciences, College of Agriculture and Environmental Sciences, Haramaya University, Dire Dawa, Ethiopia

[2]Herbal Research and Development Institute, Gopeshwar, Chamoli, Uttarakhand, India

[3]Protection of Plant varieties and Farmers' Right Authority, NASC Complex, DPS Marg, New Delhi, India

[4]School of Plant Sciences, College of Agriculture and Environmental Sciences, Haramaya University, Dire Dawa, Ethiopia

ABSTRACT

Background

Traditional and indigenous communities in India are of the religious belief that medicinal groves and plants are sacred in nature. Sacred groves (SGs) are patches of trees on forest land that are protected communally with religious zeal and connotations. These forest areas have been protected since ages by traditional societies and indigenous communities with their socio-cultural and religious practices. Sacred groves as a rule are treated piously. Sacred trees are prohibited from cutting and not axed except when wood is needed for the religious purposes like construction and repair of temple buildings or in cases like worshiping, death ceremonies and temple rituals. Thus, SGs carry direct and everlasting pious status and assist in maintaining social fabric of the society.

Results

From the present study it is concluded that, religious identification of medicinal plants and practices have influenced the folklore towards a sense of selfless services in the name the Gods. However, during the course of modernization, mechanization and globalization in the recent past has transformed and weakened both cultural and biological integrity. Changes in social belief, modernization and erosion of cultural practices are some of the major factors contributing towards degradation of the ancient institutional heritage which need to be looked into.

Conclusion

The present study is, therefore intended to propose management and conservation as an alternative strategy towards the sustainability of forests around human settlements and also an attempt to explore the role of SGs in conservation and management of different ecosystem services.

BACKGROUND

India is among the 12 mega biodiversity countries in the world having 25 hot spots of the richest and highly endangered eco-region of the world (Myers et al. [2000]). Due to increasing pressure of population, degradation and depletion of forests, it remains a challenging task to manage forests in a sustainable manner. India is classified under one of the low forest cover countries and account for 23.8% of total geographical area of the country (FSI Forest Survey of India [2005]). This percentage is still low as highlighted in the Forest Policy ([1988]). There are about 0.63 million villages (Rehmani [2012]), out of which nearly one-third are in the vicinity of forests. Thus, quite a large population is traditionally dependent on forests for a variety of reasons such as food, shelter, fodder, timber, and medicine which are directly linked to livelihood of these societies. Conservation of natural resources including biodiversity has been an integral part of several indigenous communities in and around the forest areas. The concept of sacred groves is very old and has existed since the time of Rig-Veda (an ancient Indian sacred collection of Vedic Sanskrit hymns), when tree worship was quite popular and universal. Buddha's study and enlightenment was also in the company of a banyan tree (Negi [2005]). The concept and beliefs of sacred trees and groves of forests are one of the best practices to conserve the natural resources. Sacred groves (SGs) are small forest areas that are left untouched by the local inhabitants to be conserved by the local village folk deities and are of special spiritual significance to them (Figure 1). Sacred groves, also known as church forests, fetish forests and sacred forests, are found all over the world including Ethiopia, Japan, Morocco, India, and Ghana (Cardelús et al. [2013]). The concept of SGs has been a part of rich traditional and diverse culture of Indian societies for many generations. Sometimes, they are also known as natural museums of giant trees, treasure houses of threatened species, dispensaries of medicinal plants, regulators of water sheds, recreation centers for urban life, veritable gardens for botanists, gene banks of economic species, paradise for nature-lovers and laboratory for environmentalists (Bhagwat [2009]; Manikandan et al. [2011]). Interestingly, they are found worldwide in all part of different continents i.e. Africa, Asia, Europe, Australia and America (Hughes and Chandran [1997]). Around 1,00,000 to 1, 50,000 sacred groves are reported in India (Malhotra et al. [2007]). In India the highest number of SGs (5000) has been reported to be present in state of Himachal Pradesh followed by Kerala and Chhattisgarh. Different scientific reports revealed that these SGs can support and conserve different fauna and flora (Khan and Tripathi [2004]; Chandran et al. [1998]; Sukumaran and Jeeva [2008]; Ray and Ramachandra[2010]; Singh et al. [2010]). Indian ancestors were fully aware about that the importance and significance of natural resources that sustained them which were necessary to be conserved for the sustenance of future generations. They lived in

Figure 1: Sacred trees around temple.

In India, different religions having different traditions, beliefs, and rituals are associated with conservation of biodiversity and forests. In Hindu religion, it is a traditional belief that nature shows a reverence for five basic elements i.e., Earth (*Prithvi*), Fire (*Agni*), Water (*Jal*), Air (*Wayo*) and Space (*Akash*). All the five elements are treated as a body of God and are worshipped. These five elements are protected for religious, cultural and spiritual reasons. There are many studies entitled to further quantify this ethics, which leads to biodiversity conservation and sustainable ecosystem (Pal and Mukhopadhyay [2011]).

The SGs play an important role in ecosystem services such as clean environment i.e., air, soil and water conservation, flora and fauna conservation, carbon sequestration, temperature control and conservation of traditional knowledge. They are therefore, the epicenter of ecological conservation research and policy regarding conservation and management of forest at state and national levels (Ray and Ramachandra [2010]). There are several studies carried out by various researchers on this subject highlighting significant role and potential of the SGs. The present paper therefore, explores the role of SGs in conservation and management of different ecosystem services in Indian society.

RESULTS AND DISCUSSION

Taboos for Plant and Animal Conservation

From the current study it was found that, in Chamoli district of Uttarakhand state (India), people believed that when a village rural girl gets married, the married couple has to plant a seedling of a tree in bride's house. Planting a tree seedling is a cultural and traditional practice in this region which later gained momentum in late eighties and succeeded in creating a lot of awareness among the locals. Planting of sapling as a marks of respect, is undertaken by VIPs during their visit to various establishments. This practice has been popular and planting a tree seedling has been taken up by non-government organizations (NGOs) as an important movement (Maiti Andolan) which includes list of taboos that slowly intensified, spreading to nearby areas of the region. This is a new and novel approach for biodiversity conservation in the hilly region through plantation, where the occurrence of land slide and natural hazards are more frequent. Another example of such conservation is the alpine plants *Saussurea obvallata* and *Delphinium vestitum* locally called Dungar and Loshkar; which have been protected through local customs and beliefs by linking these plants with local deity and not allowing any one from the village or outside to pluck the flowers except the village priest during specific day and time only (Meenakshi et al. [2011]).

Taboos are the unwritten, orally transmitted traditional and social rules that regulate human behavior (Colding and Folke [1997]; Banjo et al. [2006]). In rural areas of India, many plants and animals are considered as pious and sacred religiously having imbibed cultural values among the tribal communities. The religious belief serves as an instrument of protection of those rare forest species (Pandey [2003]).

Approach towards Eco-Retreats

The devastating impact of pollution, supplemented by deforestation can only be restored by green lung area of the sacred groves which, besides providing numerous tangible products remains the storehouse of the life gas–oxygen. If these storehouses are not properly managed and conserved, the future generation will definitely be deprived off this valuable asset (Anthwal et al. [2006]). As a part of school curriculum, school children are often asked to plant saplings in school compounds to understand the importance of trees. Planting for the progamme are supplied by the local forest department for

Belief and Myth towards Soil and Water Conservation

Sacred groves play an important role in soil and water conservation. They improve the soil stability of the region and act as soil binder. Plants like vetiver grass (*Vetiveria zizanioides)* and*Eucalyptus* spp are maintained to bind the soil thereby preventing soil erosion. In India, water of different rivers are treated as holy among all sources and used in all rituals and worships. People take holy dip/bath at the confluence of two or three rivers called prayag or sangam (meaning union) in different parts of the country i.e., Devparyag, Allahabad, Nasik etc. As most of the cities and town are settled on the bank of rivers ie., Ganga, Yamuna, Brahmaputra, Godavari, Krishna, Kaveri, etc. which are, not only considered sacred but are also a source of drinking water supply cities in the country. Small natural water bodies are maintained, near SGs to take care of drinking water problem during drought. Villagers are now improving the surrounding areas of water bodies with concrete structure to conserve natural flowing Waterbodies and maintain it in a hygienic condition.

Approach towards Carbon Sequestration

Estimation of carbon stocks and stock changes in tree biomass are necessary for reporting to the United Nations Framework Convention on Climate Change (UNFCCC) and which is required for Kyoto Protocol reporting also. Very little work has been done so far for estimation of the carbon sequestration potential of the sacred groves except for few reports, where carbon sequestration potential of various species was estimated in Maharashtra (Hangarge et al. [2012]). The tree, *Terminalia bellirica* was found to be dominant of sequestrating 327.78 tonnes of carbon followed 221 tonnes by *Ficus amplissima* (Hangarge et al. [2012]). The species *Gnidia glauca* had lowest carbon sequestration potential i.e. 0.0808 tonnes. A SGs having thick vegetation possesses high carbon sequestration potential thereby contributing to reduced concentration of CO_2 in the atmosphere (Hangarge et al. [2012]).

Traditional Knowledge for Conservation and Management

Traditional societies are characterized by their close interconnection with nature and its resources. Hence, the traditional communities depend upon natural resources and biodiversity more for their livelihood (Ramakrishnan [1996]). Even now the NGOs working in the region make it a point to plant important utility plants like soap nut (*Sapindus mukorossi*), arjun (*Terminalia arjuna*) Roxb., sandal (*Santalum album*) to strengthen the SGs in different lower regions.

Conservation of Medicinal and Aromatic Plants through Local Belief

The SGs are believed to be a treasure house of medicinal and aromatic plants. Though most of the indigenous people residing near the groves are illiterate, they have scrupulously nurtured their traditional customs, rituals, ceremonies and a way of forest life through folk beliefs with great vigour. The fact that sanjeevani found in Himalayan region was used by Lord Hanuman (A Hindu deity) to resuscitate Lakshman the brother of Lord Rama is well known to all Indians. There is a need for protection of large number of medicinal plants as an important component of SGs in different parts of India which is already documented in various studies (Vartak et al. [1987]; Bhandary and Chandrasekhar [2003]; Pandit and Bhakat [2007]). In Doonagiri village of Nanda Devi Biosphere Reserve, Uttarakhand state, tribal people of Bhotiya community have an ancient practice of conserving the medicinal plants of the region for centuries. This is mainly attributed to their religious belief. They have deep faith that if someone from outside the village uproots the medicinal plants from their village, it is treated as an evil act which may bring misery of great order to the village folks. They nurture a belief that Lord Hanuman came to their village and uprooted all their medicinal plants which was the cause of untreatable diseases, which struck the people later. Thus, medicinal plant conservation is an integral part of sustainable living by these people with the nature (King-Oliver et al. [1997]). Till now medicinal plants like mint (*Mentha arvensis*), coriander (*Coriandrum sativum*) and fenugreek (*Trigonella foenum-graecum*) are planted as an important component of SGs and preserving Himalayan ecosystem.

Approach towards Animal Conservation

Almost all Hindu Gods are associated with animals, birds and creatures as their vehicles or vahanas. This concept is to promote harmony in nature to maintain ecosystem. Protection of animals and other creatures along with human beings is essential to maintain ecological balance. As a result, all categories of wild and domesticated animals were identified/attached to Gods (Table 1). For instance, Garuda (eagle) Vahanam (Vah in Sanskrit language means to carry or to transport) was symbolized with Lord Vishnu, Peacock to Muruga, Bull to Lord Shiva, Tiger to goddess Durga and Rat to Vinayaka. Thus, all gods typically or mythologically are symbolized with animals as their vahanas often called the deity mounts. The association of animals with Gods as Vahanams (vehicles) is a deliberate attempt to weave mythology for protection of animals and plants in order to maintain ecological balance. Often, God is icongraphically depicted riding animals as (or simply mounted upon) vahana. Even one of the trees, Kalpavrusha, is treated as vahana. Lord Shiva uses lion skin for his attire and as supports a reptile king cobra as an ornament. Serpents are venerated on Nagpanchami day and offered milk for maintaining the natural food chain and preserving environment of SGs. Monkeys as a part of respect to Lord Hanuman and bears were also treated as gods and remained a part of the army force of Lord Rama (Table 1).

Table 1: List of some common animals, birds & reptiles and their associated beliefs

S. No.	Scientific name	Local name	Beliefs/uses
1	*Panthera tigris*	Tiger	Used in rituals and medicine.
2	*Elephas maximus indicus*	Elephant	In Hindu mythology, during Sagar Manthan (Milk sea churning), the Gods (Angels) and the demons (Asura) churned the ocean for the elixir of life called "Amrit" (Nectar) which make them immortal. From the churning the 'navratnas' (nine jewels) surfaced. Some of these jewels were the Aeravatha (white elephant), Karpavruksha (tree), Kamadhenu (Holy cow). Therefore considered to be sacred among Hindus.
3	*Bos Taurus*	Cow	In Hindu mythology the cow is considered holy The cow is equated to one's own mother (hence the expression 'Gaumata). In Hindu mythology, Kamadhenu, the wish-fulfilling celestial cow. It is believed that cow could grant any wish for those who worship cow.
4	*Bos taurus*	Bull	Bull occupies a special place in Hindu mythology. The reverence of the animal has been one of the central themes of Hinduism, since ancient times. Bull is vehicle/ vahana of lord Shiva.
5	*Antilope cervicapra*	Black buck	It is also known as Krishna Mruga in Kannada.
6	*Macaca mulatta*	Monkey	Monkey is known to be associated in the army of Hindu God Hanuman hence considered sacred.

7	*Felis catus*	Cat	Cats are associated with fertility and the goddess of birth, Shakti.
8	*Pavo cristatus*	Peacock	Hindu mythology peacocks are associated with the goddess Saraswati (goddess of education & knowledge). The feathers thus represent qualities like: kindness, patience and good fortune.
9	*Naja naja*	Snake	The snake is commonly called ('Nag' in Hindi language) is worshipped by people across the country. Some of the snakes are considered in mythology as 'protectors' and other categories as harmful/destroyers'. Lord Shiva wore snake as an ornament. Lord Vishnu sleeps on snake with thousand heads Adiseshu (King cobra) as his bed on milk sea.
10	*Bos grunniens*	Yak	The bushy tail is in great demand as a fly–wish (fan) in many Buddhist monasteries and temples.
11	*Gyps indicus*	Indian Vulture	According to the epic Ramayana, when Ravana was abducting Sita to his kingdom of Lanka, a fierce battle took place between him and the vulture King Jatayau. The bird was fatally wounded it its attempt to rescue Sita from Ravana.
12	*Columba turica*	Dove/Pigeon	Goddess Rati, the Goddess of passion and lust and the consort of Kama, the God of love, is believed to be usually depicted with a pigeon as her vahana.
13	*Corvus splendens*	House Crow	The house crow occupies a special place in Hindu religious rituals. It is usually identified with departed souls of ancestors. In Hindu the ritual of 'pinda pradhana' (offering of cooked rice balls) to the crow.

14	*Sus scrofa*Linnaeus	Wild Boar	The third avtara of Lord Vishnu was Varaha, a boar. Varaha is generally depicted having a boars head on human body. Varahi is the Hindu Goddess Durga in the form of a wild boar.
15	*Ratufa indica*	Squirrel	Squirrels are considered sacred in India and are not to be harmed. Squirrels are association with Lord Rama. During the construction of the Adi Sethu (bridge) at Rameshwaram by Lord Rama and the vanara sena, a little squirrel also contributed in its own little way. Lord Ramas pleased by the creatures dedication, caressed the squirrels back and ever since, the Indian squirrel carried white stripes on its back, which are believed to be the mark of Lord Ramas fingers.

Kandari *et al.*

Kandari *et al.* Environmental Systems Research 2014 3:16, doi:10.1186/s40068-014-0016-8

In many traditional Hindu mythological literatures the killing or hurting a cat is a crime which is also sinful and harmful. If any person is involved in killing the cat he has to make a statue of the cat of gold/ bronze metal and kept submerged in deep water to get solace and relief from the effects of the committed sin. The scientific reasoning behind the Hindu mythology was that cat kills and eats many insects and rats as an integral part of food chain, thereby preventing the spread of many viral diseases. Hence, cats need to be protected which are human friendly with their livelihood practices. In Rajasthan (western-India), the Bishnoi community treats Black Buck Deer as their child and killing or harming them is considered a great sin. Thus, Bishnoi cult supports the wildlife protection act as Black Buck Deer is considered as human friendly in selected localities (Kala and Sharma [2010]). Such incidences are widely prevalent and practiced in other countries also. The cow worship is another example of deep religious belief in India. Cow milk is treated as source of balanced diet for children in the Indian food basket. Cow dung is treated as having anti infection properties and used for domestic purposes. Cow urine is believed to possess rich chemical content and is highly useful to treat disease of human beings and is also applied as pesticide. With all its multidimensional uses of cow products, cows are treated sacred and prohibited from killing and worshipped in the name of Kamadhenu (Table 1).

Conservation of Floral Diversity

India is among the few countries where trees or forests are also worshiped during their religious rituals. Majority of plants/trees are accepted as sacred by Hindus and hence worshipped. Sacred groves are the best example of *in-situ* conservation of biodiversity, where flowers like hibiscus (*Hibiscus rosa-sinensis*), marigold (Tagetes *erecta*), jasmine (*Jasminum officinale*) Brahmakamal (*Saussurea obvallata*) are found in plenty. These SGs harbour many endemic species some of which are at different threat levels in different regions in the country. The state of Uttarakhand is endowed with world heritage site valley of flowers, where different flowers are found in abundance near SGs. Sacred groves are the natural gene pool preserver and example of habitat preservation through community participation (Gadgil and Vartak [1975]). Scientific reports also confirm the fact that SGs protect a variety of flora and fauna (Ray and Ramachandra [2010]). Sukumaran et al. ([2008]) highlighted the species richness of 212 miniature sacred groves of Tamil Nadu. Due to this deep rooted religious system, many trees are not axed as they are supposed to increase longevity. In Uttarakhand, *Cedrus deodara* is treated as a religious tree and worshipped during rituals and religious ceremonies. In Madhya Pradesh, the Gond tribes only use the fallen parts of the tree and cutting this sacred tree is totally prohibited. The neem tree (*Azadirachta indica*) is considered to be a manifestation of Goddess Durga. Besides, Tulsi plant (*Ocimum sanctum*) is worshipped as Goddess by women throughout the country which is supposed to enhance the longevity of their husbands' life.

Socio-Cultural Functions to the Sacred Groves

Many rituals are performed in the sanctity of SGs for want of rains, good health of livestock and for fending off any impending disaster. In some parts of Karnataka, people offer miniature images of cattles to keep their livestock's healthy (Kalam [1996]). Among the Bhils of Ratanmal, food and liquor are offered to SGs on the occasion of Deepawali a national festival (Nath [1960]). Likewise, in West Bengal, social gathering takes place in the vicinity of SGs on the occasion of Salui and Karam festivals (Deb and Malhotra [1997]). Different festivals such as Holi, Deepawali, and Navratri are performed around SGs in Ratnagiri district of Maharashtra (Godbole et al. [1998]). Similarly, Anthwal et al. ([2006]) reported several festivals related to SGs in Uttarakhand. Many plants species have also been associated with religious festivals viz: *Azadirachta indica* (Sheela asthami, Nimb saptami), *Ficus bengalensis* (Vrat savitiri), *Aegle marmelos* (Bilvamengal sawan ke somvaar), *Musa paradisica* (Kadii vrat) and *Ficus religiosa* (Somvari amavasya) due to popular and common beliefs (Table 2).

Table 2: List of some sacred plants in Hindu mythology

S. No.	Scientific name	**Local name**	**Family**	**Beliefs/uses**
1	*Aegle marmelos* (L.) Corr.	Bail	Rutaceae	Ritual/medicine/ sacred
2	*Artemisia nilgirica* (Clarke)	Kunju	Compositae	Ritual
3	*Atrocarpus heterophyllus* Lamk.	Theibong	Moraceae	Ritual
4	*Azadirachta indica*	Neem	Meliaceae	Ritual/medicinal
5	*Blumea balsamifera* (L.) DC.	Bari ilaichi	Compositae	Ritual/medicine
6	*Betula utilis*	Bhoj patra	Betulaceae	Ritual/sacred
7	*Calotropis procera (Ait.) R. Br.*	Aak	Asclepiadaceae	Rituals
8	*Cedrus deodara*	Deodar	Pinaceae	Sacred
9	*Cynodon dactylon* Pers.	Dhoob	Poaceae	Ritual
10	*Datura stramonium*	Datura	Solanaceae	Ritual/medicinal
11	*Desmostachya bipinnata*	Kush	Gramineae	Ritual

12	*Elaeocarpus Ganitrus Roxb.*	Rudraksha	Eleocarpaceae	Sacred/Ritual
13	*Emblica officinalis*	Amla	Phyllanthaceae	Sacred/medicinal
14	*Erythrina indica* Lam.	Paribhadraka	Faboideae	Ritual
15	*Ficus benghalensis*	Bargad	Moraceae	Ritual/sacred
16	*Ficus religiosa*	Peepal	Moraceae	Sacred
17	*Mangifera indica* L.	Mango	Anacardiaceae	Ritual
18	*Musa paradisica*	Banana	Musaceae	Ritual
19	*Ocimum sanctum* L.	Tulsi	Lamiaceae	Sacred/medicinal
20	*Pinus roxburghii*	Chir	Pinaceae	Ritual
21	*Princepia utilis*	Bhakel	Rosaceae	Ritual
22	*Prosopis cineraria*	Khachri	Leguminosae	Rituals
23	*Prunus cerasoides*	Paiya	Rosaceae	Rituals
24	*Punica granatum*	Daru	Lythraceae	Ritual
25	*Santalum album*	Sandal	Santalaceae	Ritual
26	*Saussurea obvallata*	BramKamal	Asteraceae	Sacred
27	*Terminalia arjuna* (Roxb.)	Arjun	Combretaceae	Ritual/medicine
28	*Toona ciliata* M. Roem.	Tun	Meliaceae	Ritual
29	*Xanthoxylum achanothopodum*	Timoor	Rutaceae	Sacred/Rituals/ medicine

30	*Zanthoxylum armatum*	Timoor	Rutaceae	Ritual/medicine
31	*Quercus leucotrichophora*	Oak	Fagaceae	Sacred

Kandari *et al.*

Kandari *et al. Environmental Systems Research* 2014 3:16, doi:10.1186/s40068-014-0016-8

Threats to the Sacred Groves

SGs are showing sign of weakening and decrease in terms of cultural, biological and ecological integrity in different states of the country. The number of SGs in different states also depends upon the magnitude of these threats which is area specific (Table 3). Over the past few decades many people have lost their rights of forest management to the government. Many scholars have worked on conservation of sacred groves through socio-cultural practices in different parts of India (Gadgil and Vartak [1975]; Boojh and Ramakrishnan [1983]; Khiewtam and Ramakrishnan [1989]; Rodgers [1994]; King-Oliver et al. [1997]; Tiwari et al. [1998]; Sinha and Maikhuri [1998]; Sunitha and Rao [1999]; Basu [2000]; Kushalapa et al. [2001]; Bhagwat et al. [2005]; Jaryan et al.[2010]; Kala and Sharma [2010]; Anthwal et al. [2010]). There are several reasons for this decrease in terms of area and number, primary being lack of documentation as a result of which it becomes difficult to conserve and maintain them. No legislative protection has been implemented so far in India.

Table 3: Total Sacred groves distribution in India (Adopted from Malhotra, 1998 and Malhotra et al., 2001)

Sr. No	State	Number of SGs	Generic name	Reference
1	Andhra Pradesh	750	*	WWF (World Wildlife Fund) Andhra Pradesh ([1996])
2	West Bengal	670	Bankura	Deb et al. ([1997])
3	Uttaranchal	1	Bugyal, Devban	
4	Tamil Nadu	448	Kovilkadu/ Nandavana	Amrithalingam, [1998]
5	Sikkim	56	Gumpa forest area	Chatterjee et al. [2000]
6	Rajasthan	9	Orans Kenkris	Jha et al. [1998]
7	Orissa	322	Jahera/ Thakuramma	Malhotra and Das [1997]
8	Meghalaya	79	Ka Law Kyntang	Tiwari et al. [1998]
9	Manipur	365	Gamkhap, Umanglai	Devi, [2000]; Khumbongmayum et al. [2004]
10	Maharashtra	1600	Devrai, Devrahati	Deshmukh et al. [1998]
11	Kerala	2000	Sarpakkavu/ Nagavanam	Rajendraprasad [1996]
12	Karnataka	1214	Devarabana/ Sidharavanam	Kalam [1996]
13	Jharkhand	21	Sarana/Jaherthan	
14	Chhattisgarh	2000	Devgude, Malgudi	
15	Gujarat	29	*	Gupta et al. [2000]
16	Haryana	248	*	NAEB National Afforestation and Eco-development Board [1995]
17	Himachal Pradesh	5000	Dev ban	Sharma [2000]
18	Assam	40	Miadaico	Deb et al. [1997]

19	Arunachal Pradesh	58	Gumpa Forest	Chatterjee et al. [2000]
Total		13720		

**Generic names are unavailable.*

Kandari *et al.*

Kandari *et al.* Environmental Systems Research 2014 3:16, doi:10.1186/s40068-014-0016-8

Lack of awareness in terms of long term future benefits has also resulted in the destruction of SGs. Various religious beliefs, influx of large number of pilgrims and tourists and conversion to other religions axing of plants for monetary benefit have contributed consistently for the degradation of these well flourished areas of vegetation to a mere crunch of trees in many parts of India. Greed for certain plants like sandal (*Santalum album*) has resulted in decrease of their numbers in India and particularly in Himalayan SGs. The indigenous species found in the sacred groves are threatened by the introduction of exotic weeds such as Common Floss flower (*Eupatorium odoratum*), Shrub verbenas (*Verbena officinalis*), Lantana (*Lantana camara*) and Parthenium (*Parthenium hysterophorus*) etc. Likewise, grazing, lopping and removal of biomass have also resulted in dwindling of the groves. Encroachments of SGs areas by local communities or various other government departments, migration and immigration of people have contributed to the extinction of SGs. The most recent threat to SGs is witnessed from the process of modernization, industrialization, greed of land mafias and construction of building structures resulting is loss of cultural and ecological importance among the younger generation of local people.

For providing necessary protection to the SGs and maintain their natural identity and sanctity it is imperative that cooperation of the people is extended. The surrounding village communities need to be educated and guided for sustaining the sanctity of existing grove and strengthening them. Legislation by the state government against cutting of such important trees need to be in place as is existing against cutting of khair (*Acacia catechu)* in state of Himachal Pradesh, where severe punishment is awarded for violation of the directions. During contact with woman in the villages adjoining different SGs, it was felt that they were more enthusiastic regarding conservation and preservation of different plants. Their concern was of rapid loss of important plants of utility in recent years and felt that the following issues need to be addressed without fail (Table 4).

Table 4: Issues concerns for SGs conservations

S. No	Concerns for conservation
1	Capacity building of woman regarding importance of SGs land planting by forest department for immediate multiplication of such plants for release in different villages
2	Appointing contact persons in villages on monthly basis if not on regular basis as in case of water carriers in school for maintenance of identified SGs specially near village and place of worship.
3	Involvement of school children for creating awareness and augmentation of SGs
4	Regular visits by forest staff for guidance and maintenance of SGs.
5	Scientific management of water bodies around SGs

Kandari *et al.*

Kandari *et al. Environmental Systems Research* 2014 3:16, doi:10.1186/s40068-014-0016-8

CONCLUSIONS

In the words of Dr. M.S. Swaminathan, "Unlike a botanical garden, where a wide range of trees and plants are collected and cultivated for the purpose of our education and enjoyment, the sacred groves are one method of expressing the gratitude of human beings to the trees which sustain and support life under a given agro-ecological condition". We therefore need immediate measures to stop the destruction of SGs and start their conservation as being an integral part of the social and cultural life in rural India. Documentation of all the sacred groves needs to be taken up on high priority basis so that management and conservation programs for these threatened groves can be initiated. The bondage between people, their beliefs and rituals associated with SGs in the past have hidden scientific truth. Many ancient trees are surviving till date due to the sacred belief and worship. However, in the past heavy inflow of tourists around sacred groves, threatened their very existence. Loss in faith and religious conversions in general has further intensified the magnitude of erosion of beliefs and value system. Hence, revival of SGs institution for the conservation of valuable biodiversity and ecosystem services is necessary. At the same time the cultural, biological, social and ecological importance of SGs cannot be ignored. Therefore, there is a great need for restoration and management of SGs to preserve scientific ecological system. A scientific understanding of harmony and co-existence of flora and fauna will strengthen the importance of SGs and play an important role in designing strategies for their rehabilitation and management for maintenance of sustainable ecological balance. Awareness building regarding the importance of sacred groves is first and foremost step needed to be taken for the revitalization of these traditional values, practices and beliefs among the youth. Benefit sharing should be promoted within the local communities for conserving and maintaining the groves. Stringent legislation followed by punishment accordingly is definitely needed for long and healthy life of sacred groves and therefore conservation as a whole.

Some Positive Signs

The Social Forestry wing of the Kerala Forest Department has signed memorandums of association with the owners of select groves for protection, enhancement of biodiversity and awareness programmes and 15 sacred groves have received assistance for their conservation efforts (The Hindu, May 5, 2013, India)

Chilkigarh a sacred grove being located almost in trijunction of the three states (West Bengal, Bihar and Orissa) is known for its rich and unique floral wealth was indiscriminately exploited earlier and presently the entire sacred complex has been taken over by the local Kanak Durga Temple trust consisting of members from Chilkigarh and adjoining villages. They have revived the concept of social fencing through inculcating the traditional socio-religious and ecological values of plants and simultaneously getting economic benefits through the tourists, visitors and devotees.

METHODS

A good number of SGs are found scattered in the hills of Garhwal and Kumaon region (Uttarakhand) India near the bank of rivers, Bhagirathi, Mandakani, Alaknanda as well as in the world famous spiritual centers of Yamunotri, Gangotri, Kedarnath and Badrinath. The places are visited by pilgrims in thousands, throughout the year. To collect information for the present study, a small team was constituted to visit such areas with a prepared questionnaire on different type of plants, utilized by people in adjoining villages, extinct plants and their use for religious ceremonies and awareness of these plants as well as common animals & birds attached to associated beliefs. Ten centers each were visited in the areas of Uttarkashi, Pauri, Tehri and Chamoli for this purpose. Necessary information was collected from five people in each centre majority of whom were functioning priests serving in the area. Thus, a total of 200 people were contacted for their views on the present status of the SGs. The information was later complied and tabulated.

AUTHORS' CONTRIBUTIONS

LSK contributed in the experiment design, data collection and manuscript writing and editing. VKB, MB and AKT helps in manuscript writing and data collection. All authors read and approved the final manuscript.

ACKNOWLEDGEMENTS

Authors are thankful to Haramaya University, Dire-Dawa, Ethiopia for providing necessary facilities. Contribution and help from traditional medicine practioners (Vaidyas) and other local healers of the study area in India are hereby acknowledged.

REFERENCES

1. Amrithalingam M (1998) Sthala Vrikshas of Tamil Nadu. In: Krishna N, Prabhakaran J (eds) The Ecological Traditions of Tamil Nadu, C.P.R. Environment Education Centre, Chennai.
2. Anthwal A, Sharma RC, Sharma A (2006) Sacred groves: Traditional way to conserving plant diversity in Garhwal Himalaya, Uttaranchal. The J Am Sci 2(2):35-43
3. Anthwal A, Gupta N, Sharma A, Anthwal S, Kim KH (2010) Conserving biodiversity through traditional beliefs in sacred groves in Uttarakhand Himalaya, India. Resour Conserv Recy 54:962-997
4. Banjo AD, Otufale GA, Abatan OL, Banjo EA (2006) Taboo as a means of plant and animal conservation in South-Western Nigeria: a case study of Ogbe River and its basin. World Appl Sci J 1:39-43
5. Basu R (2000) Studies on sacred groves and taboos in purulia district of west Bengal. Indian For 1:309-1317
6. Bhagwat SA (2009) Ecosystem services and sacred natural sites: reconciling material and non-material values in nature conservation. Environ Values 18:417-427
7. Bhagwat SA, Kushalappa CG, Williams PH, Brown ND (2005) A landscape approach to biodiversity conservation of sacred groves in the western Ghats of India. Conserv Biol 19:1853-1862
8. Bhandary MJ, Chandrasekhar KR (2003) Sacred groves of dakshina kanada and udupi districts of Karnataka. Curr Sci 85:655-1656
9. Boojh R, Ramakrishnan PS (1983). In: Strategies for Environmental Management, Souvenir Vol. Dept. of Science and Environmental of Uttar Pradesh, Lucknow. pp 6-8
10. Cardelús CL, Scull P, Hair J, Baimas-George M, Lowman MD, Eshet AW (2013) A preliminary assessment of Ethiopian sacred grove status at the landscape and ecosystem scales. Divers 5:320-334
11. Chandran MDS, Gadgil M, Hugues JD (1998) Sacred groves of the Western Ghats of India. In: Ramakrishnan PS, Saxena KG, Chandrashekar UM (eds) Conserving the Sacred for Biodiversity Management, Oxford and IBH Publishing Co.Pvt Ltd, New Delhi.

12. Chatterjee S, Sastry ARK, Roy BN, Lahon R (2000) Sacred groves of Sikkim and Arunachal Pradesh. Abstract. National Workshop on Community Strategies on the Management of Natural Resources, Bhopal.
13. Colding J, Folke C (1997) The relations among threatened species, their protection, and taboos. Conserv Ecol 1:1-6
14. Deb D, Malhotra KC (1997) Interface between biodiversity and tribal cultural heritage: an exploratory study. J Hum Ecol 8:157-163
15. Deb D, Deuti K, Malhotra KC (1997) Sacred grove relics as bird refugia. Curr Sci 73:815417
16. Deshmukh S, Gogate MG, Gupta AK (1998) Sacred groves and biological diversity: providing new dimensions to conservation issue. In: Ramakrishnan PS, Saxena KG, Chandrashekhara UM (eds) Conserving the Sacred for Biodiversity Management, Oxford and IBH, New Delhi. pp 397-414
17. Devi S (2000) Sacred groves of Manipur. Abstract. National Workshop on Community Strategies on the Management of Natural Resources, Bhopal.
18. (1988)
19. (2005) State of Forest Report 2005. Forest Survey of India, Government of India, Dehradun, India.
20. Gadgil M, Vartak D (1975) Sacred groves of India- a plea for continued conservation. J Bombay Natu Hist Soc 72:313-320
21. Godbole A, Watv A, Prabhu S, Sarnaik J (1998) Role of sacred groves in biodiversity conservation with 1ocal people's participation: A case.study from Ratnagiri district, Maharashtra. In: Ramakrishnan PS, Saxena KG, Chandrashekara UM (eds) Conserving the Sacred for Biodiversity Management, Oxford and IBH Publishing Co., New Delhi. pp 233-246
22. Gupta A, Shukla S, Koradiya D, Bhavsar P, Ramji A, Patel P, Taviyad R (2000) A cultural and ecological study of sacred groves in Balaram Ambaji and Jessore sanctuary in Banaskantha district of Gujarat. Abstract. National Workshop on Community Strategies on the Management of Natural Resources, Bhopal.
23. Hangarge LM, Kulkarni DK, Gaikwad VB, Mahajan DM, Chaudhari N (2012) Carbon Sequestration potential of tree

species in Somjaichi Rai (Sacred grove) at Nandghur village, in Bhor region of Pune District, Maharashtra State, India. Ann Biol Res 3(7):3426-3429

24. Hughes DJ, Chandran SMD (1997) Paper Presented in the Workshop on the Role of Sacred Groves in Conservation and Management of Biological Resources. KFRI, Peechi.
25. Jaryan V, Uniyal Gopichand SK, Singh RD, Lal B, Kumar A, Sharma V (2010) Role of traditional conservation practice: highlighting the importance of Shivbari sacred grove in biodiversity conservation. Environmentalist 30:101-110
26. Jha M, Vardhan H, Chatterjee S, Kumar K, Sastry ARK (1998) Status of Orans (Sacred groves) in Peepasar and Khejarli villages in Rajasthan. In: Ramakrishnan PS, Saxena KG, Chandrashekara UM (eds) Conserving the Sacred for Biodiversity Management, UNESCO and Oxford IBH Publishing, New Delhi. pp 263-275
27. Kala M, Sharma A (2010) Traditional Indian beliefs: a key towards sustainable living. Environmentalist 30:85-89
28. Kalam MA (1996) Sacred Groves in Kodagu District of Karnataka (South India): A Socio-Historical Study. Institute Francais de Pondicherry, Pondicherry, India.
29. Khan ML, Tripathi RS (2004) Sacred groves of Manipur- ideal centers for biodiversity conservation. Curr Sci 87(4):430-433
30. Khiewtam RS, Ramakrishnan PS (1989) Socio-cultural studies at the sacred groves at Cherrapunji and adjoining areas in north-eastern India. Man India 69(1):64-71
31. Khumbongmayum AD, Khan ML, Tripathi RS (2004) Sacred groves of Manipur: ideal centres for biodiversity conservation. Curr Sci 87(4):430-433
32. King-Oliver IED, Chitra V, Narasimha D (1997) Sacred groves: traditional ecological heritage. Int J Ecol Environ Sci 23:463-470
33. Kushalapa CG, Bhagwat SA, Kushalapa KA (2001) Conservation and management of sacred groves of Hodagu, Karnataka, South India-a unique approach. In: Ganeshaiah KN, Shaanker UR, Bawa KS (eds) Tropical Ecosystems: Structure, Diversity and Human Welfare. Oxford IBH Publishing Co. Pvt. Ltd, New Delhi. pp 565-569

34. Malhotra KC (1998) Anthropological dimensions of sacred groves in India: An overview. In: Ramakrishnan PS, Saxena KG, Chandrashekhara UM (eds) Conserving the Sacred for Biodiversity Management, Oxford and IBH Publishing Co. Pvt. Ltd.,,. pp 423-438

35. Malhotra KC, Das K (1997) Interface between faunal biodiversity and cultural heritage in south-west Bengal, in India. In: Fujiki N, Macer RJ (eds) Bioethics in Asia, Eubois Ethics Institute, Japan. pp 346-351

36. Malhotra KC, Gokhale Y, Chatterjee S, Srivastava S (2001) Cultural and Ecological Dimensions of Sacred Groves in India. Indian National Science Academy, Indira Gandhi Rashtriya Manav Sangrahalaya, New Delhi, Bhopal.

37. Malhotra KC, Gokhale Y, Chatterjee S, Srivastava S (2007) Sacred Groves in India. Aryan Books International, New Delhi, India.

38. Manikandan P, Venkatash DR, Muthuchelian K (2011) Conservation and management of sacred groves in theni district, Tamil Nadu, India. J Biosc Res 2(2):76-80

39. Meenakshi B, Chauhan NS, Kak A (2011) Dye yielding plants of Himachal Pradesh. J Econo Taxon Bot 35(2):429-432

40. Myers N, Mittermeir RA, Mittermeir CG, da Fonseca GAB, Kents J (2000) Biodiversity hotspots for conservation priorities. Nature 403:853-858

41. (1995) Sacred groves of Kurukshetra, Haryana. National Afforestation and Ecodevelopment Board, Ministry of Environment and Forests. Govt. 'of India, New Delhi India.

42. Nath YVS (1960) Bhils of Ratanmal, Maharaja Sayajirao University. Baroda. Oxford and IBH Publishing Co., New Delhi.

43. Negi CS (2005) Religion and biodiversity conservation: not a mere analogy. Int J Biodiv Sci Manage 1:85-96

44. Pal T, Mukhopadhyay M (2011) Change in Socio-cultural dimensions and its impact on exiting Sacred grove. Int J Human Soc Sci 1(10):242-250

45. Pandey DN (2003) Sacred Forest: The Case of Rajasthan. India in Forest Service, India.

46. Pandit PK, Bhakat RK (2007) Conservation of biodiversity and ethnic culture through sacred groves in Midnapore district, West Bengal, India. Indian For 133:323-344
47. Rajendraprasad M (1996) The Floristic, Structural and Functional Analysis of Sacred Groves of Kerala. Ph.D Thesis. University of Kerala, Thiriuvanthapuram, India.
48. Ramakrishnan PS (1996) Conserving the sacred from the species to landscape. Nat Resour 32:11-19 UNESCO
49. Ray R, Ramachandra TV (2010) Small sacred groves in local landscape: are they really for conservation. Curr Sci 98(9):1178-1180
50. Rehmani C (2012) Rural India, the emerging growth centre for Indian retailers. Indian Streams Res J 2(11):1-3
51. Rodgers WA (1994) The sacred groves of Meghalaya. Man India 74:339-348
52. Sharma BR (2000) Sacred groves and their role in social life in Himachal Himalayas. Abstract National Workshops on Community Strategies on the Management of Natural Resources. Bhopal, India.
53. Singh H, Husain T, Agnihotri P (2010) Haat kali sacred grove, Central Himalaya, Uttarakhand. Curr Sci 98(3):290
54. Sinha B, Maikhuri RK (1998) Conservation through socio-cultural-religious practices in Garhwal Himalayas. A case study of Hariyali sacred forest. In: Ramakrishnan PS, Saxena KG, Chandrashekara UM (eds), Conserving the Sacred for Biodiversity Management, .
55. Sukumaran S, Jeeva S (2008) A floristic study on miniature sacred forests at Agastheeshwaram, southern peninsular India. Eur Asian J Biosci 2:66-78
56. Sukumaran S, Jeeva S, Raj ADS, Kannan D (2008) Floristic diversity, conservation status and economic value of miniature sacred groves in Kanyakumari district, Tamil Nadu, Southern Peninsular India. Turkish J Bot 32:185-199
57. Sunitha S, Rao RPB (1999) Sacred groves in Kurnool District, Andhra Pradesh. In: Sivadasan M, Mathew P (eds) Biodiversity, Taxonomy and Conservation of Flowering Plants, Mentor books, Calicut. pp 367-373

58. Tiwari BK, Barik SK, Tripathi RS, et al. (1998) Sacred groves of Meghalaya. In: Ramakrishnan PS (ed) Conserving the Sacred for Biodiversity Management, Oxford & IBH Publishing Co., New Delhi, India. pp 253-262

59. Vartak VD, Kumbhojkar MS, Nipuge DS (1987) Sacred groves in tribal areas of Western Ghats: treasure trove of medicinal plants. B Medi Ethno Bot Res 8:77-78

60. (1996) Sacred Groves of Andhra Pradesh. World Wide Fund for Nature. Andhra Pradesh state Office, Hyderabad.

Chapter 7

Organic Manure Production from Cow Dung and Biogas Plant Slurry by Vermicomposting under Field Conditions

Anoop Yadav[1], Renuka Gupta[2] and Vinod Kumar Garg[1]

[1]Department of Environmental Science and Engineering, Guru Jambheshwar University of Science and Technology, Hisar 125001, India

[2]Department of Humanities and Applied Science, YMCA University of Science & Technology, Faridabad - 121006 (Haryana), India

ABSTRACT

Background

Vermicomposting is a biological process which may be a future technology for the management of animal excreta. This study was undertaken to produce vermicompost from cow dung and biogas plant slurry under field conditions. To achieve the objectives, two vermicomposting units containing cow dung (CD) and biogas plant slurry (BPS) were established, inoculated with *Eisenia fetida*species of earthworm and allowed to be vermicomposted for 3 months.

Results

After 3 months, the vermicompost was harvested and characterized. The results showed that the vermicompost had lower pH, total organic carbon (TOC), organic matter (OM) and carbon/nitrogen ratio (C/N ratio) but higher electrical conductivity (EC), nitrogen, phosphorous and potassium (NPK) content than the raw substrate. The heavy metal content in vermicomposts was higher than raw substrates.

Conclusions

During vermicomposting, the CD and BPS were converted into a homogeneous, odourless and stabilized humus-like material. This experiment demonstrates that vermicomposting is an environmentally sustainable method for the management of animal excreta.

INTRODUCTION

Vermicomposting is as a bio-oxidative process in which earthworms interact intensively with microorganisms and other fauna within the decomposer community, accelerating the stabilization of organic matter and modifying its physical and biochemical properties. The action of the earthworms in this process is physical or mechanical. Physical participation in degrading organic substrates results in fragmentation, thereby increasing the surface area of action, turnover and aeration. On the other hand, biochemical changes in the degradation of organic matter are carried out by microorganisms through enzymatic digestion, enrichment by nitrogen excrement and transport of inorganic and organic materials. The benefits of vermicomposting in recycling of organic wastes, *viz.* animal wastes (Edwards et al. 1998; Aira et al. 2002; Loh et al. 2005; Molina et al. 2013), crop residues (Bansal and Kapoor 2000), industrial wastes (Elvira et al. 1998; Kaushik and Garg 2003; Yadav and Garg 2010; Garg et al. 2012), etc., have been reported at laboratory scale, and very little attention has been given to undertake such studies in field conditions to know the success of the method.

The selection of suitable earthworm species for vermicomposting is an important step of the overall process. Out of the thousands of species of earthworms, only a few are suitable for vermicomposting of organic wastes. The epigeic species of earthworms are widely used for vermicomposting of different organic wastes. *Eisenia fetida* is the most widely used earthworm species for vermicomposting due to its well-established potential for the vermicomposting of compostable organic materials such as agricultural wastes and animal manures (Edwards et al.1998). Neuhauser et al. (1988) studied the overall reproductive capabilities of five earthworm species, *viz. Eudrilus eugeniae, Perionyx excavatus, E. fetida, Drawida veneta* and *Perionyx hawayana*, and suggested that *E. fetida* is the most appropriate species for the vermicomposting process.

Animal wastes are important resources that are used to supplement organic matters and improve soil conditions (Garg et al. 2005). Average-size cattle produces 4 to 6 tonnes of fresh dung per year. In India and some other developing nations, a significant fraction of cattle dung is used as cooking and heating fuel after making its bricks (Figure 1).

Figure 1: Cattle dung bricks prepared for cooking and heating fuel.

However, burning of dung cakes causes serious health and environmental problems. The burning of animal dung for heating and cooking results in higher indoor particle concentrations Smoke from animal dung-based cooking stoves contains carbon monoxide, fine particulates, nitrogen dioxide and hydrocarbons, all at concentrations far in excess of what is considered unsafe in outdoor air. Long-term exposure to airborne particulate matter has been associated with increased rates of acute respiratory infections, chronic obstructive lung disease and cancer (USEPA 2008). On the other hand, once the dung is burnt, the nutrients present in it are lost forever.

Another important use of cattle dung is its conversion into compost to be used as manure in agricultural fields. For this purpose, cattle dung is heaped in the open and allowed to degrade naturally without any amendments. This process takes 6 to 9 months, and still dung is not stabilized which may attract termites in the fields after its application. Further, this method is a major cause of odour and fly problems in rural areas (Figure 2).

Figure 2: Natural composting of animal dung in rural areas.

In the last few decades, the Government of India has promoted biogas production at individual and community levels using cattle dung and various other wastes. Biogas is used as cooking and heating fuel in rural areas in India. During this process, animal dung is converted into slurry which is good-quality manure and can be applied in agricultural fields as soil conditioner (Figure 3). However, it was observed that farmers are reluctant to apply it directly to agricultural fields. So some new method for pollution-free management of cattle dung and biogas plant slurry is desired.

Figure 3: Biogas plant slurry disposed in the open.

A bibliographic survey has indicated that vermicomposting can be an efficient technology for the management of CD and BPS. Various studies conducted on vermicomposting of animal excreta are encapsulated in Table 1. The available data indicate that most of the studies have been taken up at laboratory scale. There is paucity of data on the applicability of vermicomposting technology for the vermiconversion of animal wastes under natural environmental conditions. So this work was undertaken to study the vermicomposting of cattle dung and biogas plant slurry under field conditions.

Table 1: Various studies conducted on vermicomposting of animal excreta/ waste

Animal waste	Earthworm	Scale	Reference
Cow dung	*Perionyx ceylanensis*	Laboratory	Karmegam and Daniel (2009)
Cow, sheep, pig and chicken wastes	*Eudrilus eugeniae*	Laboratory	Coulibaly and Zoro Bi (2010)
Cattle manure	*Eisenia andrei*	Laboratory	Lazcano et al. (2008)
Cattle manure	*Eisenia andrei*	Pilot	Elvira et al. (1998)
Cow manure	*Eisenia fetida*	Laboratory	Contreras-Ramos et al. (2005)
Cattle solids, pig solids, horse solids and turkey waste	*Perionyx excavatus*	Laboratory	Edwards et al. (1998)
Cow manure and poultry droppings	*Metaphire posthuma*	Laboratory	Bisht et al. (2007)
Cow, buffalo, horse, donkey, sheep, goat and camel wastes	*Eisenia fetida*	Laboratory	Garg et al. (2005)
Cow manure	*Eisenia fetida*	Industrial	Aira et al. (2011)
Pig manure	*Eisenia fetida*	Laboratory	Aira et al. (2007)
Rabbit manure	*Eisenia fetida*	Laboratory	Molina et al. (2013)
Rabbit manure	*Eisenia fetida*	Industrial	Gómez-Brandón et al. (2013)
Goat manure	*Eisenia fetida*	Laboratory	Loh et al. (2005)

Yadav *et al.*

Yadav *et al. International Journal Of Recycling of Organic Waste in Agriculture* 2013 2:21, doi:10.1186/2251-7715-2-21

METHODS

Waste Collection and Earthworms (*E. fetida*)

Fresh cow dung (CD) was procured from an intensively live-stocked farm at Hisar, India. Anaerobically digested biogas plant slurry (BPS) was procured from a post-methanation storage tank of an on-farm biogas plant. The raw material used in the biogas plant was cow dung. The average nitrogen, phosphorous and potassium (NPK) content and other important parameters of cow dung and biogas plant slurry compiled from other studies are given in Table 2.

Table 2: Average range of nutrient content of CD and BPS

Serial number	Parameters	CD	**BPS**
1	pH	7.5 to 8.3	7.2 to 8.1
2	EC (dS/m)	1.2 to 2.2	1.0 to 1.4
3	TOC (g/kg)	425 to 550	400 to 470
4	TKN (g/kg)	6.5 to 8.6	5.2 to 10
5	TAP (g/kg)	5.0 to 8.7	5.0 to 5.8
6	TK (g/kg)	2.8 to 7.8	1.3 to 4.2
7	TNa (g/kg)	1.38 to 5.8	1.9 to 2.8
8	TCa (g/kg)	1.87 to 2.0	1.2 to 3.0
9	Fe (mg/kg)	1,134 to 1,884	900 to 1,200
10	Cu (mg/kg)	31 to 234	20 to 50
11	Zn (mg/kg)	110 to 143	20 to 75
12	Cd (mg/kg)	2.10 to 4.5	1.0 to 10

Yadav *et al.*

Yadav *et al. International Journal Of Recycling of Organic Waste in Agriculture* 2013 2:21, doi:10.1186/2251-7715-2-21

E. fetida earthworm species was used in the experiment. Healthy unclitellated hatchlings weighing 200 to 250 mg live weight were randomly picked up for the experiment from other vermicomposting beds maintained under field conditions.

Experimental Setup

Ten kilograms of each waste (CD or BPS) was filled in circular plastic containers of 45-cm diameter and 25-cm depth without any bedding. All vermicomposting units were established in triplicate. After 3 weeks of pre-composting, 250 unclitellated *E. fetida* hatchlings were introduced in each vermicomposting unit. All the vermicomposting units were maintained at a farmer's field without any temperature control in a thatched house. The moisture content was maintained at 70% ± 10% by periodic sprinkling of canal water throughout the study period. The ambient temperature during the study period is given in Figure 4.

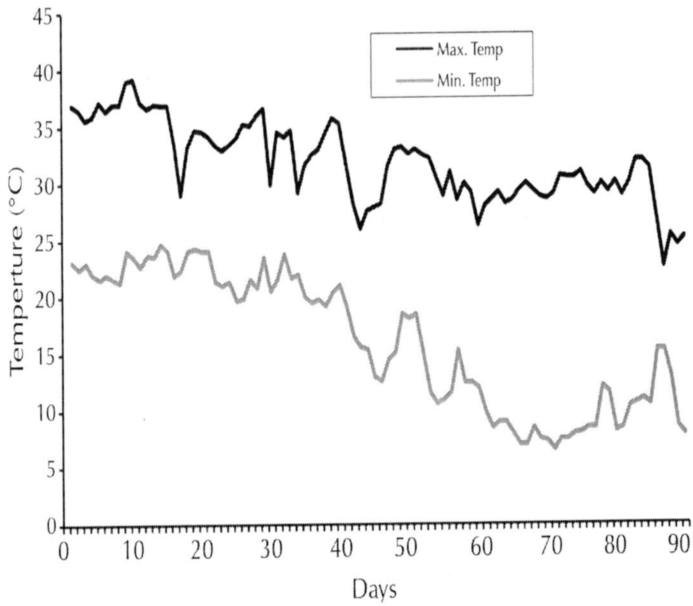

Figure 4: Minimum and maximum ambient air temperature at the experimental site during the study period.

Chemical Analysis

Homogenized samples of CD and BPS were drawn at day 0 (initial) and day 90 (at the end). Day 0 refers to the day of inoculation of earthworms after pre-composting for 3 weeks. The raw wastes and vermicomposts were air-dried at room temperature prior to physico-chemical and heavy metal analysis. The pH and electrical conductivity (EC) were determined using a water suspension of the raw waste or vermicompost in the ratio of 1:10 (*w/v*) after agitating for 30 min using pH and electrical conductivity meters, respectively. Total organic carbon (TOC) was determined as reported by Nelson and Sommers (1982). Total Kjeldahl nitrogen (TKN) was determined after digesting the sample with concentrated H_2SO_4 and $HClO_4$ (9:1, *v/v*) according to the Bremner and Mulvaney (1982) procedure. Total available phosphorus (TAP) was determined spectrophotometrically with molybdenum in sulphuric acid. Total potassium (TK) and total calcium (TCa) were determined after digesting the sample in a diacid mixture of concentrated HNO_3 and concentrated $HClO_4$ (9:1, *v/v*) using a flame photometer. Heavy metals (Fe, Cu, Zn and Cr) were determined using an atomic absorption spectrophotometer (AAS) after digestion of the sample with a diacid mixture of concentrated HNO_3 and concentrated $HClO_4$ (9:1, *v/v*). The carbon/nitrogen ratio (C/N) ratio was calculated from the measured value of TOC and TKN. The reported results are the mean of three replicates with standard deviation (mean ± SD). The probability levels used for statistical significance were $P < 0.05$ for the tests. Statistical analysis of the data was carried out with the SPSS 16.0 software program.

RESULTS AND DISCUSSION

Physico-chemical Characteristics of Vermicomposts

The physico-chemical characteristics of the raw materials and vermicomposts are presented in Table 3. The results showed that the pH values of vermicomposts were lower than those of their respective raw materials, i.e. CD and BPS (Table 3). Decreases in pH were significantly different after vermicomposting in both mixtures ($P < 0.05$). The pH reduction may be attributed to mineralization of nitrogen and phosphorus into nitrites/nitrates and orthophosphates and bioconversion of the organic material into intermediate species of organic acids (Ndegwa et al.2000).

Table 3: Changes in physico-chemical characteristics of CD, BPS and their vermicomposts (mean ± SD, *n* = 3)

Treatments	Parameters			
	pH	**EC (dS/m)**	**TCa (g/kg)**	**TK (g/kg)**
CD	7.9 ± 0.40 b	1.2 ± 0.10 a	2.0 ± 0.12 b	8.8 ± 0.20 b
CD-vermicompost	6.7 ± 0.10 a	1.6 ± 0.35 a	3.10 ± 0.15 d	9.7 ± 0.20 d
BPS	7.8 ± 0.35 b	1.4 ± 0.10 a	1.4 ± 0.10 a	4.2 ± 0.10 a
BPS-vermicompost	7.3 ± 0.20 ab	1.7 ± 0.12 a	2.45 ± ?0.18 c	6.3 ± 0.24 c

Different letters in each column represent significant differences, and the same letters do not represent significant differences ($P < 0.05$).

Yadav *et al.*

Yadav *et al. International Journal Of Recycling of Organic Waste in Agriculture* 2013 2:21, doi:10.1186/2251-7715-2-21

The EC of CD-vermicompost was 33.3% higher than that of CD, and that of BPS-vermicompost was 21.4% higher than that of BPS. The increase in EC is an indication of mineralization of wastes (Neuhauser et al. 1988). The differences in EC in the end products were insignificant ($P < 0.05$). Calcium (Ca) content in the vermicomposts was higher than that in CD and BPS. Maximum increment was observed in CD-vermicompost (3.10 g/kg) and minimum increment in BPS-vermicompost (2.45 g/kg) (Table 3). Spiers et al. (1986) have reported that earthworms convert calcium oxalate crystals in ingested fungal hyphae to calcium bicarbonate which is then egested in cast material, which increases calcium availability in the final vermicompost. Garg and Kaushik (2005) have also reported an increase in calcium content during the vermicomposting of industrial wastes. The earthworms drive the mineralization process and convert a fraction of calcium from binding form to free forms, resulting in its enrichment in worm casts.

The TK content was higher in vermicomposts than in raw wastes (Table 3). The initial TK content in CD and BPS was 4.2 and 8.8 g/kg, respectively, whereas the TK content in the BPS-vermicompost and CD-vermicompost was 6.30 ± 0.24 and 9.70 ± 0.20 g/kg, respectively. The differences in TK content in the end products obtained from different waste mixtures were significant ($P < 0.05$). Suthar (2007) has reported that an earthworm-processed waste material contains a higher concentration of exchangeable K due to enhanced microbial activity during the vermicomposting process, which consequently enhances the rate of mineralization. The production of acid during decomposition of organic matter by the microorganisms is the key mechanism for solubilization of insoluble TK (Adi and Noor 2009).

After vermicomposting, the TP content was 10.2 ± 0.40 and 7.0 ± 0.20 g/kg in CD-vermicompost and BPS-vermicompost, respectively (Table 4). The TP contents in both vermicomposts were significantly different from each other ($P < 0.05$). Le Bayon and Binet (2006) have reported that some amount of phosphorus is converted to more available forms partly by earthworm gut enzymes, i.e. acid phosphatases and alkaline phosphatases. Sangwan et al. (2010) have reported a 1.3- to

1.5-fold increase in phosphorus content after vermicomposting of press mud.

Table 4: Biochemical characteristics of CD, BPS and their vermicomposts (mean ± SD, n = 3)

Treatments	Parameters			
	TP (g/kg)	**TOC (g/kg)**	**TKN (g/kg)**	**C/N ratio**
CD	5.0 ± 0.30 a	427 ± 12 c	8.5 ± 0.33 a	50.24 ± 0.34 c
CD-vermicompost	10.2 ± 0.20 c	275 ± 16 a	28.7 ± 1.10 b	9.58 ± 0.15 a
BPS	5.6 ± 0.42 a	471 ± 33 c	6.2 ± ?0.25 a	75.97 ± 1.70 d
BPS-vermicompost	7.0 ± 0.40 b	364 ± 14 b	29.9 ± 2.10 b	12.34 ± 0.50 b

Different letters in each column represent significant differences, and the same letters do not represent significant differences ($P < 0.05$).

Yadav *et al.*

Yadav *et al. International Journal Of Recycling of Organic Waste in Agriculture* 2013 2:21, doi:10.1186/2251-7715-2-21

The TOC content of the vermicomposts was lesser than initial levels. The reduction in TOC content in vermicomposts was 35.6% and 22.7% in CD-vermicompost and BPS-vermicompost, respectively (Table 4). Aira et al. (2007) reported that earthworms modified substrate conditions, which subsequently enhanced carbon losses from the substrates through microbial respiration in the form of CO_2. Nitrogen content was higher in vermicomposts than in CD and BPS. The TKN content of BPS and CD was in the range of 6.2 and 8.5 g/kg (Table 4). The TKN content increased in the range of 28.7 to 29.9 g/kg after vermicomposting. There was a 3.37- and 4.82-fold increase in TKN after vermicomposting CD and BPS, respectively. The TKN contents in CD-vermicompost and BPS-vermicompost were not significantly different from each other ($P < 0.05$). Kaushik and Garg (2004) have reported a 2.0- to 3.2-fold increase in TKN during vermicomposting of textile mill sludge mixed with cow dung and wheat straw. Benitez et al. (1999) have reported that decomposition of organic materials by earthworm accelerates the N mineralization process and subsequently changes the N profile of the substrate.

The initial C/N ratio was in the range of 50.2 and 75.9 (at day 0) of CD and BPS, respectively, but after 90 days, there was a significant decrease in the C/N ratios (Table 4). The final C/N ratio of CD-vermicompost and BPS-vermicompost was 9.58 ± 0.3 and 12.2 ± 0.7, respectively. The reduction in C/N ratio was 80.9% to 83.9% during the vermicomposting process. The C/N ratio indicates the degree of stabilization of a waste as carbon is lost as CO_2 and nitrogen content is increased during the vermicomposting process, and these factors contribute to the lowering of the C/N ratio. The decrease in C/N ratio and relative increase in the TKN of vermicomposts may also be due to the loss of dry mass in terms of CO_2 (Viel et al. 1987). So a high degree of organic matter stabilization in CD and BPS was achieved after vermicomposting which proves that *E. fetida*can promote decomposition and mineralization of organic matter.

Heavy Metals in Vermicomposts

All vermicomposts contain trace elements and other metals which play an important role in their production, worm growth and application in agricultural fields. The potential risk of heavy metals is their transfer from the soil to food grains after vermicompost application in agricultural fields. However, the bioavailability of heavy metals to crops is influenced by several factors including the nature of heavy metals, soil properties (soil type, pH, cation exchange capacity (CEC), nutrient status, organic matter content, redox potential, texture and characteristics of the deposition/composition), plant species, agronomic conditions, etc. Heavy metal contents in raw wastes and final vermicomposts are given in Table 5. While comparing the metal concentration of the vermicomposts with initial levels, it was observed that the concentration of all the studied metals was higher in vermicomposts than initial raw materials. However, the metal concentrations in vermicomposts obtained from both raw materials (CD and BPS) were lesser than the prescribed limits for heavy metal application in agricultural soils. Total iron (Fe) content of vermicomposts was 2.3% and 3.2% higher in CD-vermicompost and BPS-vermicompost, respectively. Total copper (Cu) content was also higher after vermicomposting (14.1% to 76%). Cu content was more in BPS-vermicompost. Total zinc (Zn) content was 5.4% and 46.7% higher in CD-vermicompost and BPS-vermicompost than in CD and BPS, respectively. Chromium content was also higher in vermicomposts than in raw materials; it was 98 ± 6.8 mg/kg in CD-vermicompost and 133 ± 8.5 mg/kg in BPS-vermicompost (Table 5). The results indicate that metal content in the vermicomposted material was closely related to the metal concentration in raw materials. Gupta and Garg (2008) have reported an increase in heavy metal concentration in vermicomposts of sewage sludge after vermicomposting. Suthar and Singh (2008) have also reported a higher concentration of metals in earthworm casts collected from sewage soils and cultivated lands. The weight and volume reduction due to mineralization and decomposition of organic matter during vermicomposting may be the major factor for increase in metal concentrations in vermicomposts (Domínguez 2004).

Table 5: Heavy metal (mg/kg) content in CD, BPS and their vermicomposts (mean ± SD, *n* = 3)

Parameters	CD	CD-vermicompost	BPS	BPS-vermicompost	EU limit range [a]
Fe (mg/kg)	1,859 ± 19	1,902 ± 48	1,077 ± 21	1,112 ± 64	-
Cu (mg/kg)	234 ± 15	267 ± 30	22.5 ± 3	39.6 ± 5	70 to 600
Zn (mg/kg)	110 ± 9	116 ± 3.4	31 ± 2.4	45.5 ± 5.1	210 to 4,000
Cr (mg/kg)	84 ± 3.8	98 ± 6.8	127 ± 8.2	133 ± 8.5	70 to 200

[a]Source: Brinton (2000).

Yadav *et al.*

Yadav *et al. International Journal Of Recycling of Organic Waste in Agriculture* 2013 2:21, doi:10.1186/2251-7715-2-21

CONCLUSIONS

The results indicate that CD and BPS can be used as a raw material in vermicomposting. The vermicomposts were rich in NPK and their C/N ratio was below 15 which indicate their agronomic importance as soil conditioner and manure. The quality of raw materials determined the physico-chemical characteristics of vermicomposts. The results suggest that vermicompost can be introduced as one of the technologies for converting organic wastes into value-added products.

AUTHORS' CONTRIBUTIONS

AY carried out the experimentation and physico-chemical analysis and drafts the manuscript. RG performed the statistical and physico-chemical analysis. VKG carried out planned this study, supervision on the data analysis and revised the manuscript. All authors read and approved the final manuscript.

ACKNOWLEDGEMENTS

One of the authors (Anoop Yadav) is thankful to CSIR, New Delhi (India), for providing financial assistance in the form of the Extended Senior Research Fellowship to conduct this work.

REFERENCES

1. Adi A, Noor Z (2009) Waste recycling: utilization of coffee grounds and kitchen waste in vermicomposting. Bioresour Technol 100(2):1027-1030
2. Aira M, Gómez-Brandón M, González-Porto P, Domínguez J (2011) Selective reduction of the pathogenic load of cow manure in an industrial-scale continuous-feeding vermireactor. Bioresour Technol 102(20):9633-9637
3. Aira M, Monroy F, Domínguez J (2007) Earthworms strongly modify microbial biomass and activity triggering enzymatic activities during vermicomposting independently of the application rates of pig slurry. Sci Total Environ 385:252-261
4. Aira M, Monroy F, Dominguez J, Mato S (2002) How earthworm density affects microbial biomass and activity in pig manure. European J of Soil Biology 38:7-10
5. Bansal S, Kapoor KK (2000) Vermicomposting of crop residues and cattle dung with *Eisenia foetida*. Bioresour Technol 73:95-98
6. Benitez E, Nogales R, Elvira C, Masciandaro G, Ceccanti B (1999) Enzyme activities as indicators of the stabilization of sewage sludge composting with *Eisenia foetida*. Bioresour Technol 67:297-303
7. Bisht R, Pandey H, Bharti D, Bisht SPS, Kaushal BR (2007) Reproductive potential of the earthworm *Metaphire posthuma* (Oligochaeta) in different food substrates. Tropical Ecol 48(1):107-114

8. Bremner JM, Mulvaney RG (1982) Nitrogen total. In: Page AL, Miller RH, Keeney DR (eds) Method of soil analysis, American Society of Agronomy, Madison. pp 575-624
9. Brinton WF (2000) Compost quality standards and guidelines. Final report. Woods End Laboratories, Mount Vernon. (Accessed on 12 Feb 2013)
10. Contreras-Ramos SM, Escamilla-Silva EM, Dendooven L (2005) Vermicomposting of biosolids with cow manure and oat straw. Biol Fertil Soils 41:190-198
11. Coulibaly SS, Zoro Bi IA (2010) Influence of animal wastes on growth and reproduction of the African earthworm species *Eudrilus eugeniae* (Oligochaeta). European J Soil Biol 46:225-229
12. Domínguez J (2004) State-of-the-art and new perspectives on vermicomposting research. In: Edwards CA (ed) Earthworm ecology, 2nd edn. CRC Press, Boca Raton. pp 401-425
13. Edwards CA, Dominguez J, Neuhauser EF (1998) Growth and reproduction of *Parionyx excavatus* (Perr.) (Megascolecidae) as factors in organic waste management. Biol Fertil Soils 27:155-161
14. Elvira C, Sampedro L, Benítez E, Nogales R (1998) Vermicomposting of sludges from paper mill and dairy industries with *Eisenia andrei*: a pilot-scale study. Bioresour Technol 63:205-211
15. Garg VK, Chand S, Chhillar A, Yadav A (2005) Growth and reproduction of *Eisenia foetida* in various animal wastes during vermicomposting. Appl Ecol Environ Res 3(2):51-59
16. Garg VK, Kaushik P (2005) Vermistabilization of textile mill sludge spiked with poultry droppings by an epigeic earthworm *Eisenia foetida*. Bioresour Technol 96:1063-1071
17. Garg VK, Suthar S, Yadav A (2012) Management of food industry waste employing vermicomposting technolog. Bioresour Technol 126:437-444
18. Gómez-Brandón M, Lores M, Domínguez J (2013) Changes in chemical and microbiological properties of rabbit manure in a continuous-feeding vermicomposting system. Bioresour Technol 128:310-316
19. Gupta R, Garg VK (2008) Stabilization of primary sewage sludge during vermicomposting. J Hazard Mater 162:430-439

20. Karmegam N, Daniel T (2009) Growth, reproductive biology and life cycle of the vermicomposting earthworm, *Perionyx ceylanensis* Mich. (Oligochaeta: Megascolecidae). Bioresour Technol 100:4790-4796

21. Kaushik P, Garg VK (2003) Vermicomposting of mixed solid textile mill sludge and cow dung with epigeic earthworm *Eisenia fetida*. Bioresour Technol 90:311-316

22. Kaushik P, Garg VK (2004) Dynamics of biological and chemical parameters during vermicomposting of solid textile mill sludge mixed with cow dung and agricultural residues. Bioresour Technol 94:203-209

23. Lazcano C, Gómez-Brandón M, Domínguez J (2008) Comparison of the effectiveness of composting and vermicomposting for the biological stabilization of cattle manure. Chemosphere 72:1013-1019

24. Le Bayon RC, Binet F (2006) Earthworm changes the distribution and availability of phosphorus in organic substrates. Soil Biol Biochem 38:235-246

25. Loh TC, Lee YC, Liang JB, Tan D (2005) Vermicomposting of cattle and goat manures by*Eisenia foetida* and their growth and reproduction performance. Bioresour Technol 96:11-114

26. Molina MJ, Soriano MD, Ingelmo F, Llinares J (2013) Stabilisation of sewage sludge and vinasse bio-wastes by vermicomposting with rabbit manure using *Eisenia fetida*. Bioresour Technol 137:88-97

27. Ndegwa PM, Thompson SA, Das KC (2000) Effects of stocking density and feeding rate on vermicomposting of biosolids. Bioresour Technol 71:5-12

28. Neuhauser EF, Loehr RC, Malecki MR (1988) The potential of earthworms for managing sewage sludge. In: Edwards CA, Neuhauser EF (eds) Earthworms in waste and environmental management, SPB Academic Publishing, The Hague. pp 9-20

29. Nelson DW, Sommers LE (1982) Total carbon and organic carbon and organic matter. In: Page AL, Miller RH, Keeney DR (eds) Method of soil analysis, American Society of Agronomy, Madison. pp 539-579

30. Sangwan P, Kaushik CP, Garg VK (2010) Vermicomposting of sugar industry waste (press mud) mixed with cow dung employing an epigeic earthworm *Eisenia fetida*. Waste Manage Res 28:71-75
31. Spiers GA, Gagnon D, Nason GE, Packee EC, Lousier JD (1986) Effects and importance of indigenous earthworms on decomposition and nutrient cycling in coastal forest ecosystems. Canadian J Forest Res 16:983-989
32. Suthar S (2007) Nutrient changes and biodynamics of epigeic earthworm *Perionyx excavatus* (Perrier) during recycling of some agriculture wastes. Bioresour Technol 98(8):1608-1614
33. Suthar S, Singh S (2008) Feasibility of vermicomposting in biostabilization of sludge from a distillery industry. Science of Total Environment 394(23):237-243
34. USEPA (2008) National Ambient Air Quality Standards. US Environmental Protection Agency. http://www.epa.gov/air/criteria.html (Accessed on 15 Feb 2013)
35. Viel M, Sayag D, Andre L (1987) Optimization of agricultural, industrial waste management through in-vessel composting. In: De Bertoldi M (ed) Compost: production, quality and use, Elsevier, Essex. pp 230-237
36. Yadav A, Garg VK (2010) Bioconversion of food industry sludge into value-added product (vermicompost) using epigeic earthworm *Eisenia fetida*. World Revi Sci Technol Sust Dev 7(3):225-238

Chapter 8

The Dairy Chains in North Africa (Algeria, Morocco and Tunisia): From Self Sufficiency Options to Food Dependency?

Mohamed Taher Sraïri[1], Mohammed Tahar Benyoucef[2], and Khemais Kraiem[3]

[1]Department of Animal Production & Biotechnology, Hassan II Agronomy and Veterinary Medicine Institute, Madinate Al Irfane 10 101, Rabat, Morocco

[2]Department of Animal Productions, National High School of Agronomy, El Harrach, Algiers, Algeria

[3]Department of Animal Resources, Food Technology and Rural Development, Higher Institute of Agricultural Science, Chott Mariam, Tunisia

ABSTRACT

The Maghreb countries (Algeria, Morocco and Tunisia) have experienced since the early 1950s a rapid demographic growth coupled to a significant rhythm of urbanization. This has led to a marked increase in the demand of dairy products. In order to secure the supply, specific policies have been implemented. They mainly consisted in the establishment of a dairy industry, based on the processing of either raw milk produced locally (in Morocco and Tunisia) or imported milk powder (in Algeria). These divergent options have had significant consequences on the whole organization of the dairy chains in these countries, from cattle rearing practices, to milk collection and processing. They have also implied differences in milk and its derivatives' prices and levels of consumption. The paper draws a comparative analysis of milk chains within the three countries: a supply mainly based on imports in Algeria, whereas in Morocco and Tunisia, the demand is satisfied by a chain relying on locally produced cattle milk. The paper also emphasizes on the future challenges that will have to be addressed: a rising volatility of milk and other strategic inputs' prices (feed, machinery, cattle, etc.) in global markets, an improvement in consumers' awareness about milk quality, a further pressure on natural resources (mainly soils and water) to get more raw milk, in countries already suffering an acute water stress. The article also establishes recommendations about specific issues related to the development of the dairy chains in the context of North Africa. These are mainly linked to the fragmented offer induced by numerous smallholder farms, which implies obvious difficulties to assess the hygienic and the chemical quality of milk batches delivered daily. Moreover, this fragmented offer also means that specific support programs will have to be designed, as the vast majority of farms are not dairy specialized, expecting both milk and calf crop from their herds.

INTRODUCTION

Located in the Northern fringes of Africa, the Maghreb countries (Algeria, Morocco and Tunisia) have a long tradition with dairy products' consumption. In fact, their original dietary habits used to be based on cereals and vegetables, and on a limited intake of animal products, mainly *leben*(fermented milk), which all contributed to a low prevalence of obesity and cardiovascular diseases (Mehio Sibai et al. 2010). Moreover, according to the Berber and the Arabic traditions, milk has a symbolic value of life and fertility, as it is often used, with dates, in ceremonies to welcome guests (Benchelah and Maka 2008). The three countries have experienced since the Independence era (early 1960s) a rapid demographic growth coupled to changes in the diet. As a consequence, a massive surge in food demand has occurred (Lampietti et al. 2011), particularly in animal products (meat and milk), implying, as in many other developing countries the need for a "Livestock Revolution" (Delgado et al. 2003). Therefore, specific policies had to be implemented to secure the supply of food and create wealth and labor opportunities through the use of the natural resources: arable and range lands, animal wealth and irrigation water. In the specific field of animal products, dairying has been intensively supported since the colonial era, given its better metabolic efficiency in comparison to meat production (Vermorel and Coulon 1998). Several policies were therefore applied in each country, in relation to their resources availability and the weight of the rural areas in their domestic affairs. Schematically, Algeria with its prominent fossil energy exports, relied mainly on imports of milk powder to provide its citizens with subsidized dairy products (Henry2004), becoming nowadays one of the most important dairy products' importer worldwide. At the opposite, Morocco and Tunisia made different choices, based on encouraging locally produced raw milk. Such options have had important consequences on the whole organization of the dairy chains within the three countries. Given that these policies have been established for more than 40 years, it is now possible to draw a comparative diagnosis of the three dairy chains. In fact, many social, economic and environment challenges remain associated to cattle rearing in North Africa, as it is a key activity to ensure food security. It induces however marked impacts on the environment, given its important needs of water (Sraïri et al. 2009a) and its effects on groundwater pollution by nitrates (Laftouhi et al. 2003). Moreover, the current volatility in the prices of feed and food, and its effects on the imports' bill put additional pressure on the actors of the dairy chains within the three countries, as their intrinsic water stress and the expected climate change already threaten the prevailing agricultural activities (Schilling et al. 2012). The current study aims to show the most recent trends and future prospects of the dairy chains in Algeria, Morocco and Tunisia. It also attempts to investigate the drivers for the development of milk production, consumption, and trade in the region, and the constraints that will have to be confronted.

The Context of Dairy Production in the Maghreb Region

History

Livestock rearing is an ancient activity in the Maghreb region. In fact, this area is characterized by an original population of native cattle which are locally known as the Brown Atlas breed. There is strong evidence that this breed has many morphological similarities with the sub Saharan N'dama breed, as pastoral tribes used to cross the Sahara desert in attempts to find feed and water for their herds (Smith 1992). Another evidence of cattle herding activities in the region is the presence of rock art, disseminated in many parts of the Maghreb area, representing scenes of cows' milking (Dunne et al. 2012).

Milk production has remained for many centuries practised by pastoral societies. It is the need to intensify cattle production, as a consequence of the demographic growth of the twentieth century, that has induced the emergence of fodder crops to feed cattle, at a time where livestock feeding was almost based only on range land resources, cereal crops' by-products (straw, stubble and bran), in addition to fallow (Davis 2007). Recent history in the Maghreb region is also characterized by a rapid demographic expansion and by growing rhythms of urbanization. The official figures reveal that the population has more than doubled from 1970 to 2012: 13.7 to 37.1 millions in Algeria, 14.9 to 32.7 millions in Morocco and 5.1 to 10.7 millions in Tunisia (World Bank 2012). This trend of population growth is expected to continue until 2050, though these countries have entered a phase of demographic transition at different paces. Further, the overall population of the three countries is expected to reach around 105 million people in year 2050 (Population Reference Bureau 2010). Coupled to urbanization rates currently exceeding 60%, as they were inferior to 15% in the beginning of the 20th century, this has implied a marked evolution in the consumption habits, with more meals eaten alone, outside home. This trend has induced a surge in the demand of animal products with emerging risks of a food dependency for the region (Sraïri 2011).

Geography and Natural Resources

The Maghreb countries are mainly characterised by a typical Mediterranean climate, with a long summer period (from May to the end of September) of intense drought and excessive heat, followed by erratic rainfall from autumn to spring (October to April) (Lionello et al. 2006). Another factor which affects the Maghreb climate is the marine influence, as it decreases the amplitude of temperatures in areas near to the seashore: the Atlantic Ocean in Morocco, and the Mediterranean in Morocco, Algeria and Tunisia. The average annual rainfall is well below 300 mm in vast areas of the Maghreb countries, implying arid to semi arid climates (Patricola and Cook 2010). Therefore water stress constitutes the main limiting factor to agriculture. The hydrological water stress index which measures hundreds of people per one million cubic meters of available renewable water is respectively 29, 11 and 23 in Algeria, Morocco and Tunisia respectively (World Resources Institute2008). It implies that at the regional level, Algeria and Tunisia face the highest level of water stress, while in Morocco water is less scarce. Such figures will certainly worsen with the expected demographic growth and climate change and they may threaten a harmonious human development (Rijsberman 2006). The relative better figure in Morocco is related to the Atlas mountains which have favoured the implementation of a policy of dams and large scale irrigation schemes named the "1 million irrigated hectares" (Faysse et al. 2010).

In addition to irrigation policies, the agricultural output in the Maghreb remains largely related to the level of annual rainfall in rain fed areas (Thomas 2008), with no possibilities of irrigation and which are nutrients' shortages for ruminants prone. In fact, in the three countries, the availability of arable land per capita remains limited (respectively 0.24, 0.29 and 0.49 ha in Algeria, Morocco and Tunisia), and irrigation possibilities are reduced (Table 1). With regard to milk production, the contribution of the rain fed areas is highly variable and there is an ongoing trend towards localizing the dairy activity in the irrigated basins, where water availability enables to alleviate the effects of drought. For example, in Morocco, the irrigated regions which represent less than 15% of the total arable land contribute to up to 60% of the total annual output of milk (MAPM 2011).

Table 1: Key statistics of the Maghreb countries

	Algeria	**Morocco**	
Country area (10^3 ha)	238,174	71,085	16,215
Agricultural area (10^3 ha)	8,459	9,232	5,045
Irrigated area (10^3 ha)	907	1,454	345
Population in 2012 (× 1,000)	37,120	32,678	10,673
Urban population (%)	63	55	65
Life expectancy	72	74	75
GDP per capita - US $* (2009)	7,740	4,108	7,520
Contribution of agriculture to GDP (%)	8.3	17.1	10.6

*Purchasing power parity.

Source: CIA, 2010 and UNDP, 2009.

Sraïri et al.

Sraïri et al. SpringerPlus 2013 2:162, doi:10.1186/2193-1801-2-162

Animal Resources

Another important characteristic of the dairy production in the Maghreb area is the reduced contribution of non-cattle species to the overall output. In fact, because of the harsh environments which prevail in many parts of the region, small ruminants (sheep and goat) flocks are mainly used for meat production. The existing breeds of these species have not been selected for milk yield, at the exception of the Sicilo Sarde nucleus in Tunisia (Atti et al. 2006), where the vicinity with Sicily and the Ottoman colonization have induced an old tradition of cheese consumption made from ewe milk. Thus, the official figures reveal that milk from non cattle species (small ruminants and camel) represents respectively 21.3, 5.1 and 3.7% of the overall output in Algeria, Morocco and Tunisia (FAO STAT 2012), and its industrial processing remains rather weak. Therefore, integrated dairy chains rely mainly on cattle milk.

Cattle numbers vary significantly among the Maghreb countries: respectively 1.6, 2.8 and 0.6 millions in Algeria, Morocco and Tunisia. These figures showed very slight evolutions in recent years (Figure 1), as the official policies in the three countries encourage an improvement of the average milk yield per cow rather than the increase in the number of cattle. To achieve such an increase of the milk output, one of the most prominent measures adopted was a program of crossbreeding of local strains with high genetic merit breeds like the Holstein, the Montbéliarde and the Brown Swiss. Consequently, imports of cattle of such breeds were encouraged. In Morocco, such imports are nowadays subsidized to support the constitution of a local population of such breeds. As a consequence, the imports of pregnant heifers have reached since the early 1960s respectively some 387,000, 350,000 and 80,000 in Algeria, Morocco and Tunisia. In addition to cattle imports, the Maghreb countries have all implemented programs of artificial insemination (AI), using semen of high genetic merit dairy cattle. The official figures reveal that the number of AI reached in year 2011 some 204,600; 320,000 and 305,000 in Algeria, Morocco and Tunisia. Though the number of AI is increasing steadily, its efficiency could be significantly improved as the conception rate in herds is frequently superior to 2 (Sraïri and Farit 2001). All together, these measures have implied a stable increase in the population of crossbred and purebred dairy cattle, at the expense of local breeds, which show a limited dairy potential (less than 1,000 kg per lactation). It is estimated that purebred dairy cows which used to represent less than 10% of the total cattle population in the early 1970s have increased to 17, 23 and 55%, respectively in Algeria, Morocco and Tunisia. The differences may be explained by the dairy policies adopted. In Algeria, milk powder imports have constituted a significant constraint for the local development of raw milk production and collection. They have also hindered the connexion between agricultural services and dairy cattle farms. Milk imports have also induced an insufficient interest in animal recording systems (Benyoucef and Abdelmoutaleb 2010). In Morocco, the constraints are related to the geography characteristics where an important population of local cattle is located in remote mountainous areas. These cattle breeding intervention measures coupled to continuous efforts to promote dairy intensification through irrigated forage, concentrates use and mechanic milking extension have prompted a continuous increase in the cattle milk output in the Maghreb countries (Figure 2).

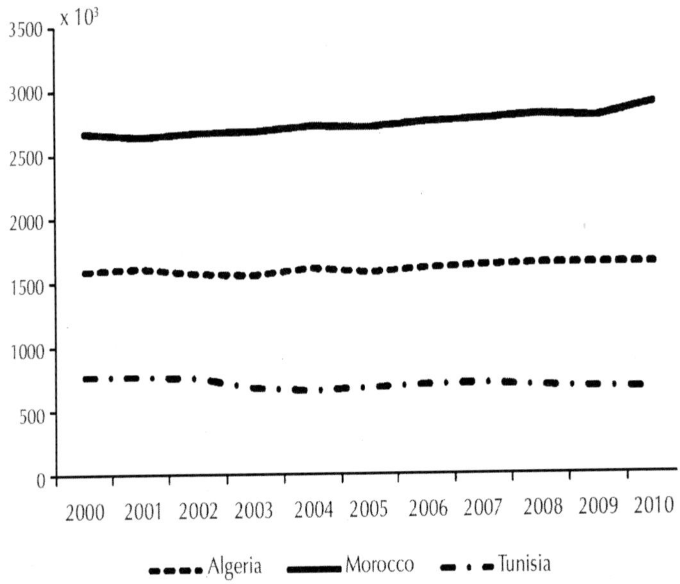

Figure 1: Recent evolutions of cattle numbers in the Maghreb countries (2000–2010).

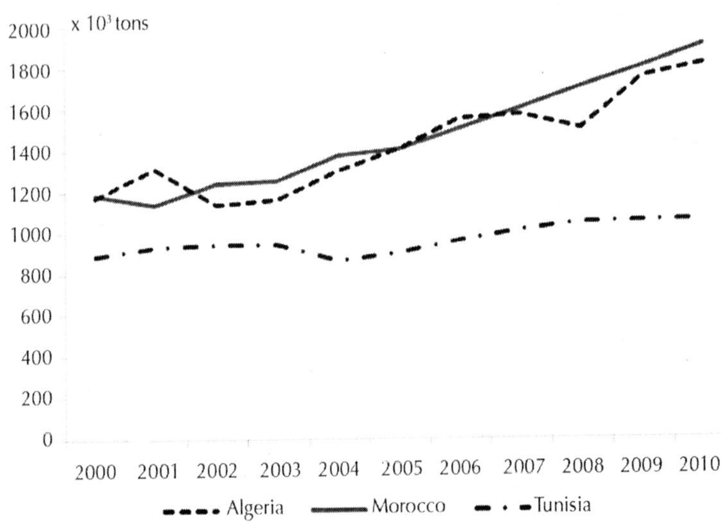

Figure 2: Recent evolutions of cattle milk output in the Maghreb countries (2000–2010).

The Dairy Policies Adopted and Their Consequences

Since the early 1960s, with the Independence era, the agricultural policies and the management of rural areas have played an important role in the development of the Maghreb countries. That is particularly true in Morocco and Tunisia, whereas in Algeria, the petroleum sector was more prominent. Therefore, in both Morocco and Tunisia, the major objective was to try to attain self-sufficiency in the most commonly consumed food products (cereal grains, milk, sugar, etc.), whereas in Algeria, the options were quite different, the authorities relying on gas exports to ensure the supply of the population with imported goods (Benyoucef 2005; Aït Amara 2009). In the particular field of the dairy chains, these options have had marked consequences.

Structure of Milk Production and Collection

The main characteristics of cattle milk production in the Maghreb region is a fragmented offer, induced by numerous smallholder farms. The official figures indicate that in Algeria, Morocco and Tunisia, farms with less than 5 cows and which rely on less than 5 ha represent almost 80% of the total number of farms delivering milk to the supply chain. Such a structure implies a relatively high number of farms with cattle (respectively 212,000; 720,000 and 112,000 in Algeria, Morocco and Tunisia), with marked differences in their production characteristics: genetic type of cattle, feeding systems, respective weight of meat and milk in the production strategies, etc. Attempts to establish typologies of farms on their functioning characteristics have revealed that variables such as the efficiency of feed resources conversion to milk, the level of dairy intensification (mainly milk yield per lactating cow) and the economic profitability are the most decisive, with no effect of farms geographic location (whether in an irrigated scheme or a rain fed suburban region) (Sraïri et al. 2009b). Therefore, cattle's feeding represents an important constraint in this region and has significant consequences on farms' performances. Thus, the climate characteristics of the Maghreb region induce marked effects for the sustainability of intensive dairy farming, because of frequent drought seasons, which also affect the irrigated areas (Sraïri and Ilham 2001), as water volumes within dams is not sufficient to support an important demand from various crops, particularly in summer time, where temperatures might reach 50°C. The main constraint is represented by frequent insufficient and imbalanced dietary rations, which alter milk yield. Many research studies have demonstrated that the average annual milk yield per cow in conventional herds, even with high genetic merit dairy breeds or crossbred cows, does not exceed 2,500 to 3,000 kg (Madani and Mouffok 2008; Rekhis et al. 2007; Sraïri et al. 2009b), whereas in Tunisia, in herds with purebred dairy cattle breeds it reaches 3,500 to 4,000 kg (GIV Lait 2012). At farm level, dairy cattle feeding systems in various areas of the Maghreb region suffer from acute limitations, as cattle load (number of animals per ha of fodder) is frequently too high. Moreover, as the vast majority of cattle farms are not dairy specialized, there is often a significant competition for feed resources between lactating cows and growing calves, which means that farmers have to make painful choices between the two functions. To overcome such limitations, farmers used to rely on two options: *i*) the important use of off-farm feed resources, through purchased feed like cereal grains (maize and barley mainly), crop by-products, such as beet pulp and bran, and proteaginous meals (soya been and sunflower) which may represent up to 60% of the total energy intake (Kadi and Djellal 2009; Sraïri and Kessab 1998), and *ii*) the incorporation of lignified poor roughages at high levels in dietary rations, given that these resources like straw or hay can easily be harvested and transported from cereal specialized regions to areas with an intensified cattle activity (Annicchiarico et al. 2005). Those two options show obvious limitations. First of all, in the current context, given the soaring prices of feed in global markets and the stagnating farm gate milk prices, the use of off-farm resources may hamper significantly the economic profitability of the dairy activity (Rejeb Gharbi et al. 2007). Second, the use of cereal straws and poor roughages (lignified oat hay) as the pillar of cattle feeding systems in dry seasons may also decrease significantly milk yield. In fact, such feeding resources show a reduced nitrogen content (less than 10%) and a limited digestibility (less than 50%) which cannot sustain the feeding requirements of high genetic merit cows (Ørskov 1999). As a consequence, milk yield in dry periods often decreases dramatically, as farmers are not well aware about the techniques of dietary rations complementation and they even often ignore cattle requirements exactly. Moreover, such feeding systems characteristics also imply frequent reproduction failures, which contribute to alter the overall farms' profitability (Sraïri and El Khattabi 2001; Madani et al. 2002). To enhance the performances, support programs for farmers, based on an effective follow-up of herds, are urgently required.

Raw milk fragmented offer in the Maghreb countries also implies that a collection infrastructure has to be established. It often consists in dairy collection co-operatives, which are generally a group of neighbouring farmers who install a refrigeration device to collect milk and aggregate it before delivering their output to dairy processing units. The evolutions and the performances of this collection infrastructure in the Maghreb countries are rather contrasted. In fact, with a dairy policy mainly relying on milk powder imports to secure the supply since the early 1970s, the development of an efficient chain using local raw milk has not been achieved yet in Algeria. The unfair competition with cheaper imported milk powder has not enabled local operators to invest in the collection network (Benyoucef, 2005). However, with the global food crisis of 2007 and its effects on milk imports' bill, the Algerian authorities have effectively activated the National Inter professional Milk Office (ONIL), as a regulation agency with the mission to organize the dairy sector. One of its main objective was to increase significantly the rate of raw milk collection, as it reached 22.4% of the national output in 2011, while this figure did not exceed an average value of 13%, from 2000 to 2010 (DRDPA 2011). The relative improvement of milk collection rate has been made effective through a regular increase in the financial incentive given to collection centres, from 0.03 to 0.06 and finally 0.09 US $ per kg of milk, respectively in 2000, 2003 and 2009. As a consequence, in year 2011, it is estimated that raw milk collection has reached the 560 million litres (DRDPA 2011). These evolutions still mean however that almost 80% of the milk output escapes from industrial processing, being consumed on-farm or sold in short scale local circuits. As an example, in some peripheral regions of the south of Algeria, the dairy chains remain embryonic, because of the absence of industrial processing units. Milk produced by dairy cattle farms outside the oases is sold directly to consumers. By contrast, milk from small ruminants' species (goats and ewes) is consumed on-farm, by the breeders' family members (Benyoucef and Boubekeur 2010).

In Morocco and Tunisia, the situation is rather different. Milk collection centres play a crucial role within the dairy supply chain. They also constitute vital operators in regional political affairs, given their interactions with local communities and their steady incomes. The number of milk collection points has been continuously increasing in both countries as it reached in year 2011, 230 in Tunisia and more than 1,450 in Morocco. They gather up to 60% of the total output in the two countries (Table 2).

Table 2: Main indicators of the dairy chains in the maghreb countries (2010)

	Algeria	Morocco	Tunisia
Number of cattle (x 1,000)	1,600	2,800	600
Pure breed dairy cattle (%)	17	23	55
Cattle milk output (× 10^3 tons)	1,811	1,900	1,059
Non cattle milk in the overall milk output (%)	21.3	5.1	3.7
Formal milk collection rate (%)	14	65	63
Number of industrial milk processing units	144	44	41
Average annual milk consumption (kg per capita)	100	55	110
Self-sufficiency rate in dairy products (%)	49	89	95

Source: FAO STAT, 2012 and national statistics compiled by the authors.

Sraïri et al.

Sraïri et al. SpringerPlus 2013 2:162, doi:10.1186/2193-1801-2-162

Raw milk collection centres' performances are generally acceptable, as their main function is to gather milk from farmers and to sell it to industrial plants. In Morocco, their gross margins are secure, as there is a differential between farm gate price which is paid to farmers and the price of milk sales to industrial processors. Generally, that differential is up to 5 to 10% of farm gate milk price, whether farmers deliver by their own means their daily production or the collection centre sends a small pick up to farms to get milk.

In Tunisia the State authorities implemented since 1988 a subsidy for milk collection centres which reached 0.025 US $ per kg in 1994 to secure a reasonable margin. Until now, the organisation of the dairy chains in both Morocco and Tunisia show that milk collection points have been successful, as they allow farmers even in remote areas to integer industrial food chains and earn a vital daily income. However, such structures show evident limitations when it comes to assessing regularly individual milk quality and rewarding it consequently (Le Gal et al. 2009).

Dairy Processing Operators in the Maghreb Countries

The operators in the industrial dairy processing sector in the Maghreb area can be classified according to three criteria: their industrial capacity (quantity of milk processed yearly), their status (totally private, co-operative, or State owned) and the type of products they sell (drink milk and/or dairy derivatives). In addition to these industrial processors, there are also numerous artisanal dairy units.

In Algeria, the leading public company is GIPLAIT, an old state group which manages some 14 units disseminated throughout the country (Bencharif 2001). It is currently undergoing restructuring with the aim to privatise some of its plants. They mainly operate with imported milk powder and milk fat. They supply the market with almost 70% of the total quantities of subsidized drink milk and 30% of the dairy derivatives, mainly yogurts and cheeses (Benyoucef 2005).

In addition to these State run units, there are 130 private units of various industrial capacities in the dairy processing sector. The vast majority of these factories (99) operate with imported milk powder to produce rather drink milk or dairy derivatives. Given the recent changes in the dairy policy in Algeria, and the emergence of a strong political will to encourage locally produced raw milk channels, some processing units have settled recently with the aim of avoiding the use of imported milk powder. There are nowadays 21 units of various capacities using raw milk to get drink milk (Ultra High Temperature - UHT) and 10 processing units which use local raw milk to produce dairy derivatives. These recent developments have also been encouraged by the intervention in the dairy processing activity of international groups like "Sodiaal" (brand name Candia) and "Danone", who are seeking joint ventures with local groups with the objective to invest in this rapidly growing supply chain, in a country recently integrating the market economy (Cheriet et al. 2008).

In Morocco, the dairy processing sector is rather different, as from its very beginning it has always been totally private. It consists in some 44 societies which may be classified in four different types.

The first one gathers private industrial operators run by independent societies. There are four main societies operating in this sector, of which the "Centrale Laitière" group, occupying a leading position with almost 55% of the volumes processed. This society has established strong ties with the international dairy group "Danone", which controls in 2012 some 67% of its capital. Its joint venture with "Danone" provides "Centrale Laitière" with two important assets: technical know-how and the ability to sell products under the well-known "Danone" label (Jazi 2003). In 2008, following the global food crisis, "Centrale Laitière" invested massively in a milk dryer to ensure domestic supply. This facility is currently the most important milk dryer in Africa. In fact, this has been a logical step for the company to avoid milk losses in times of excess supply (especially during rainy seasons in years with important precipitation). Therefore, the milk is dried and conserved until periods of milk shortage, as this also allows the company to decrease its imports of milk powder, which at the time was very expensive. Similarly, "Centrale Laitière" has launched in year 2008, following the food global crisis an important project of dairy cattle production called "Lait Plus". This intensive dairy farm aims to rear some 10,000 lactating cows in the Gharb irrigated scheme, also characterized by its average annual rainfall above 600 mm. This was an additional step adopted by the "Centrale Laitière" group to secure its supply of raw milk and to try to improve the quality of the raw matter it processes. Such a project is also supported by the ongoing trend of agricultural intensification promoted by the Moroccan authorities, known as the "Green Morocco Plan" (MAPM 2011).

In addition to "Centrale Laitière", there has been in 2005 the emergence of another important player in the private sector of dairy processors. This is the "Safilait" company, known locally by its brand "Aljibal" (Mountains in Arabic). The plant is situated in the irrigated scheme of Tadla (centre East of Morocco) where the company runs a dairy farm - Agroplus - of more than 3,000 cows of both Holstein and Montbéliarde breeds. The annual processed quantity of milk reaches some 30,000 tons representing around 4% of the national milk production. During the last 5 years, the "Aljibal" brand name has gained a reputation of good quality and its products can be found in supermarkets in large cities.

Apart from these two operators, there are also two other private smaller groups. The first one is the "Domaines Agricoles", a private company mainly retained by the Royal family. It is producing niche market dairy products under the brand "Chergui". This operator processes only milk from three of its own farms located in the North of Morocco, with a herd estimated to contain some 1,000 lactating cows. Its dairy plant processes around 9,000 tons annually, exclusively in milk derivatives, such as yogurts and *leben*, sold about 10% more expensive than the average price of conventional products of other brands.

Finally, there is also the global group "Nestlé", present in Morocco but only with some high value products such as ice creams, dairy specialities for infants and milk powder. Its dairy basin is located in the Doukkala large scale irrigation scheme where it processes around 3% of the volumes collected by industrial plants.

The second type of dairy processors is represented by a large national cooperative, called "COPAG", which is based in the Southern region of Souss-Massa. Over the past twenty years, this cooperative has become a strong competitor to "Centrale Laitière". "Copag" incorporates some 11,000 dairy farmers milking around 40,000 cows. The co-operative represents some 20% of the total milk processed and its market shares are progressively increasing, as the quality of its drink milk and dairy preparations, sold under the brand name "Jaouda" (Quality in Arabic) has been recognized. Moreover, its co-operative status lets the company benefit from certain goodwill among Moroccan consumers (Errahj et al. 2009).

Other than the large players, there is a host of smaller processing co-operatives based in a number of different dairy basins supplying only their immediate geographical markets. This part of the processing sector is very complex and highly fragmented, as it gathers some 30 co-operatives representing less than 15% of the milk volumes processed. Some of these small co-operatives were dairy processing pioneers in Morocco ("Le Bon Lait" in the Marrakech region, "Colait Extralait" in the Gharb, etc.) but they are currently struggling to survive as they face severe financial and technical problems such as increases in milk prices and other strategic inputs' costs (such as imported milk powder, butter, energy, etc.). They produce only a narrow range of dairy products that is not usually stocked by supermarkets and they only have little know-how when it comes to milk processing. Some of them, in an attempt of resilience, like "Le Bon Lait" are currently establishing a joint venture with the French leading dairy co-operative group "Sodiaal" to market the brands Candia and Yoplait in the Moroccan market.

Finally, the traditional processing sector also deserves mention. These small operators are known locally as *mahlabates*. They are mainly located in popular suburbs in large cities where they only serve a small niche market. There are however no official data specifying the exact milk volumes going through this informal sector, though it is estimated they process around 10 to 15% of Morocco's raw milk. These circuits benefit from both farmers' and consumers' acceptance. Indeed, farmers manage to sell their milk at prices 15 to 20% higher than to collection centres.

Similarly to Morocco and Algeria, the milk processing sector in Tunisia is fragmented, with many companies of different size and ownership structure. Dairy processing activities are practiced by 41 societies: 8 which produce mainly drink milk (Ultra High Temperature), 4 are specialized in yogurts, 28 manufacture cheese and 1 which processes milk to powder.

The "Delice Danone" group is a strong leader in the dairy processing activities in Tunisia. It has two different plants, one for drinking milk and the other for yogurts' production. This group represents almost 50% of the overall drink milk market shares and 70% of total sales of yogurts (Ben Mahmoud and Jemni 2008). Two other important dairy processors are Vitalait and Laino, both active in drink milk and yogurts, but with limited market shares. Tunisia has a long tradition of cheese processing and some companies are specialized in that activity. This is the case of Centrale des Produits Laitiers Souani, which represent 60% of total cheese production of the country.

Drink milk represents more than 75% of total milk processed, whereas yogurts and cheeses represent only 13 and 8% of total volumes (Figure 3). The 4% remaining are converted to dried milk, particularly in periods of high production, used in case of milk shortage.

In fact, in Morocco, as the supply is mainly based on the deliveries of local raw milk, the seasonality of production constitutes a real constraint to dairy processors. As periods of milk excess are frequent in the rainy season, it creates significant tensions to remunerate farmers. Some dairy processors have tried to overcome such problems by drying milk and using it in periods of limited production. In other situations, farmers may not be paid regularly, adding pressure on the profitability of cattle rearing (Sraïri and Chohin Kuper 2007). In Tunisia, two distinct periods of production are observed. The period of high output starts in spring and ends in the middle of summer. It is characterized by frequent milk refusal by processing units, because of its low quality (low density, high acidity, high plate counts, etc.). At the opposite, during the period of low production (from the end of summer until the end of winter), due to low forage availability, all the milk production is accepted by the processing units, regardless of its quality, because of the lack of raw matter.

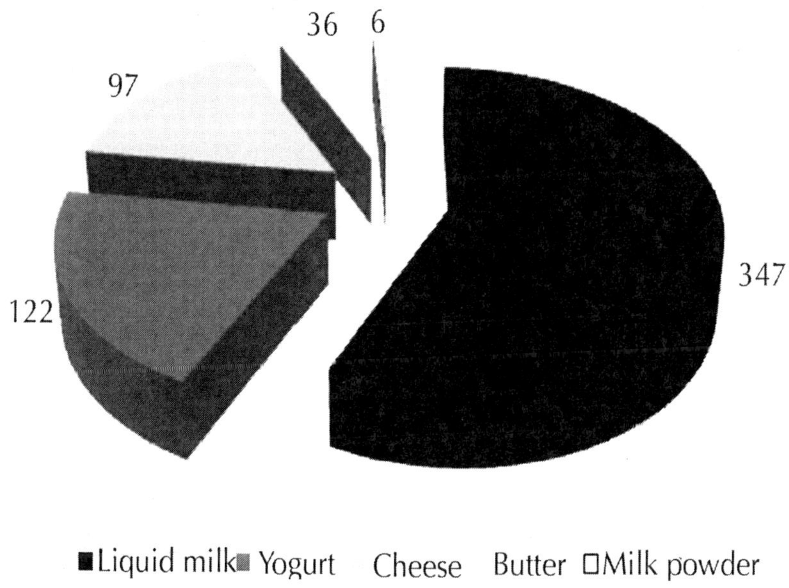

Figure 3: Split of the Tunisian dairy sector by product (average annual output -103 tons-, 2000–2009).

Milk Prices and Their Effects on Milk and Dairy Derivatives' Consumption

Milk prices and their variations according to quality criteria are crucial indicators of the dairy chains' governance (Valeeva et al. 2007). In the Maghreb countries, there are contrasted policies which have been implemented by the authorities, with regard to milk prices. In Algeria, given the aim to secure the supply mainly with imported commodities, the structure of the dairy products' market is linked primarily to the subsidized drink milk reconstituted from imported powder. This good represents up to 52% of the dairy products' overall consumption. Its price has been fixed at 25 DA per litre (0.3 US $). Such a subsidized product has been harmful to the integration of local raw milk in the supply chain, as it is estimated that it represents less than 10% of the dairy uses in the country. However, because of the soaring prices of imported milk powder and other dairy products during the global food crisis of 2007, a significant trend to increase the use of local raw milk has occurred. This has been noticed through the adoption by the State authorities in year 2008 of a series of incentives to increase raw milk production (0.14 US $ per kg), collection (0.06 US $ per kg) and industrial processing (0.07 US $ per kg) (DRDPA 2011). In addition, State authorities have adopted other incentives to promote cattle rearing (19 US $ per conception through AI, a subvention of 30% of the investments in milking machine, ensiling material, etc.) and milk collection (up to 30% of the investments in cooling devices). All these intervention measures aim to decrease the bill of dairy imports, and try to promote a sustainable supply chain based on locally produced fresh milk. As a consequence, farm gate milk price which stagnated for many years at 0.23 US $ per kg has increased in 2009 to 0.33 US $. During the same period, the price of the subsidized drink milk reconstituted from imported powder has remained administrated by the State to 0.27 US $. Given the lower price of the subsidized milk in comparison to raw milk, even before its processing, the Algerian authorities try to control the market of subsidized milk through ONIL. All together, the structure of the Algerian dairy market proves that the authorities prefer to promote mass consumption at the expense of the development of the local cattle industry. As a result, the average annual milk and dairy products' consumption reaches almost 100 kg *per capita*. Milk powder accounts for 51.5 kg per capita, followed by drink milk (37.0 kg), fermented milk (8.7 kg), cheese (1,6 kg) and finally yogurt (1,2 kg) (ONS 2010).

In Morocco, milk prices are rather different. With the structural adjustment policies applied to the Moroccan economy at the beginning of the 1980s, the liberalization of the dairy chain's governance has induced the end of State intervention. Therefore, from 1992, milk prices, whether at farm gate or at the consumer are totally free of any State intervention (Aït El Mekki and Tyner 2004). As the imports of dairy products are heavily taxed, the Moroccan market is mainly supplied with locally produced raw milk. Long term evolutions of farm gate and consumption milk prices reveal a significant trend of increase of the value perceived by dairy processors at the expense of cattle farmers (Sraïri and Chohin Kuper, 2007) (Table 3). Such an evolution has recently induced the increase of tensions within the dairy chain between farmers and dairy processors, particularly with the marked increase of major inputs' prices (above all feed). In comparison to the Algerian situation, milk price at consumption remains higher (almost 0.75 US $ per kg of drink milk), as there are no subvention specific to that product. As a result, annual milk and dairy products average consumption is lower, as it is estimated to reach in 2010 some 55 kg per capita. Studies on the structure of milk consumption show that the consumers' perception of dairy products is rather good, as they represent almost 8% of the overall food expenditures. However milk consumption suffers from the limited income of many households, particularly in rural areas (HCP,2001). Therefore, important disparities in milk and dairy products' consumption levels remain among consumers, according to their social background and the households' income (Sraïri and Karbab, 2010).

In Tunisia, milk prices have been intensively linked to the intervention of the authorities. Rejeb Gharbi et al. (2007) have in fact clearly demonstrated that without border protection, dairy production is not profitable in Tunisia even under the best favorable hypothesis about water and concentrate feed prices. Therefore, farm gate milk price has steadily increased, to ensure the sustainability of cattle farmers. In year 2007, farm gate milk price reached a value of 0.36 US $ per kg, after an increase of almost 21% as a result of several protests of farmers due to the soaring prices of inputs. In addition to the State intervention through the increase of farm fate milk price, an incentive to encourage milk collection and processing was also implemented. It reached 0.025 US $ per kg of raw milk destined to milk collection centres (GIV Lait 2012).

Milk and dairy products' consumption is estimated to reach 110 kg per capita in year 2011, and it has been steadily increasing since the 1980s, when it did not exceed 85 kg. In addition, the structure of dairy products' consumption has evolved significantly, as drink milk quantities almost stagnated, whereas cheese and butter figures experienced marked increases (respectively 600 and 57% from the 1990s levels) (Dhehibi and Gil 2003).

Table 3: Evolution of farm gate milk price and consumption milk price in Morocco - US $ (1970/2010)

Year	Farm gate milk price (1)	**Milk price at consumption (2)**	**(1)/ (2) (%)**
1970	0.06	0.12	51.4
1975	0.10	0.13	75.4
1980	0.16	0.23	68.6
1985	0.23	0.34	66.1
1990	0.29	0.45	63.9
1995	0.33	0.55	58.8
2000	0.33	0.60	54.4
2005	0.33	0.69	47.4
2010	0.33	0.71	46.9

Adapted from Sraïri and Chohin Kuper, 2007.

Sraïri et al.

Sraïri et al. SpringerPlus 2013 2:162, doi:10.1186/2193-1801-2-162

Milk and Dairy Products' Trade

The Maghreb countries are net importers of dairy products. In fact, the Algerian total dairy imports were valued at 1.2 billion US $ in 2008 (an average annual expenditure of 36 US $ *per capita*), and this value decreased to 859 million US $ in 2009, as a result of lower international commodity prices. This still accounted for 15% of the country's all food imports, second only to cereal grains. The Algerian government controls the price for drink milk reconstituted from imported milk powder delivered only by ONIL. Algeria's tariffs and duties on dairy products range from 5 to 47%. Proximity and good freight rates from Europe have always made trade with the European Union more advantageous and easier than for other countries. The structure of the

Algerian dairy products is by far dominated by milk powder (whether skimmed or whole milk dried) which almost accounts for an annual volume of 252,000 metric tons (average value from 2006 to 2009) (Figure 4a). It is followed by cheese (an average of 18,900 tons per year, from 2006 to 2009), and finally butter (an average of 8,840 tons annually during the same period).

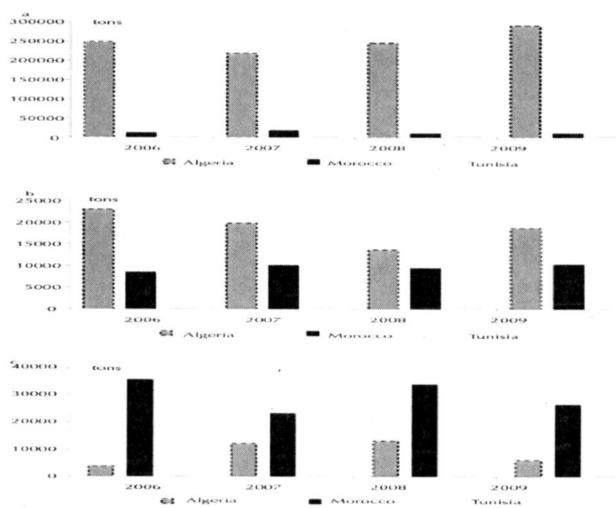

Figure 4: Dairy products imports into the Maghreb countries, 2006–2009. a Milk powder (both skimmed or whole milk dried) imports.b Cheese imports. c Butter imports.

The Moroccan dairy imports are rather different. The overall value of dairy products imports reached 227 million US $ in 2010 (Office des Changes 2012). That corresponds for year 2010 to an expenditure of less than 8 US $per capita. Morocco's tariffs/duties on dairy products range from 17.5% (cheese) to 32.5% (butter) and reaches 102% for milk powder and yogurts, in a strong sign that the authorities protect the local market, with high duties levied on dairy products.

As the domestic market is mainly supplied by locally produced raw milk, the Moroccan dairy imports are dominated by high value derivatives. Butter ranked first during year 2009 (26,600 tons) (Figure 4b), followed by milk powder and cheese, with almost the same volumes (respectively 11,500 and 10,300 tons) (Office des Changes 2012). The official data reveal that the main suppliers of dairy products to Morocco have varied significantly in recent years. Since year 2007, there has been a surge in US imports of dairy products, as a consequence of the free trade agreement signed between the two countries in 2004. Other traditional important suppliers of dairy products to Morocco are France for milk powder (9,400 tons), New Zealand and Ireland for cheese (respectively 2,300 and 2,000 tons) and New Zealand and Australia for butter (7,800 and 5,400 tons).

The Tunisian dairy imports are the most limited in the region. They did not exceed 34.3 million US $ in 2009 (3.2 US $ *per capita*). Tunisia's tariffs/duties on dairy products range from 31 to 67%: yogurt and cheese (all types) duties - 67%, crude milk, milk powder and butter duties - 57% and whey duties - 31% (Tunisian customs, 2012). Milk powder is the most imported dairy good (7,780 tons in 2009), followed by cheese (1,920 tons) and finally butter (1,400 tons) (Figure 4c).

Finally, it is worth mentioning that the Maghreb countries do not export dairy products. Some of the huge quantities of milk powder imported by Algeria are however smuggled to neighbouring Morocco, Tunisia and Libya, but the exact quantities remain unknown. In addition, Tunisia has tried to export some drink milk to Algeria, particularly in period with raw milk excess (rainy seasons). However, given the seasonality of such exports, they cannot sustain all year long the demand of the Algerian market.

Current Challenges and Future Outlook

Many challenges lie ahead for the dairy chains in the Maghreb area if they want to achieve a sustainable development and contribute efficiently to the supply of the population. One of the most important is linked to the steady increase of milk volumes in a context of scarce resources, particularly water. In fact, given the climate constraints, no dairy intensification may be possible without the production of high quality irrigated fodder. The existing figures show that on farm water productivity through cattle can be significantly improved as, from irrigation practices to fodder yield and its conversion to cattle products (milk and live weight gain), there are numerous losses. In the specific case of the Tadla large scale irrigation scheme in Morocco, with a dairy chain relying on perennial alfalfa as the main fodder in numerous smallholder farms with herds of dual purpose (milk and meat simultaneously), it appeared that almost 1.8 and 10.6 cubic meters of water were necessary to get a single kg of milk and of live weight gain respectively (Sraïri et al.2009a). Such results imply the use of huge amounts of water to ensure higher milk outputs. Unfortunately, many areas of the Maghreb countries are characterized by structural aridity and erratic rainfall which have already implied an unsustainable use of groundwater resources to intensify their agricultural activities (Wada et al. 2012).

Another important challenge for the Maghreb dairy chains is related to the complex issue of milk quality assessment and remuneration. In fact, in the intermediate collection structures, which gather milk deliveries from farmers before supplying the industrial process units, the quality of individual batches can hardly be analyzed due to technical, economic and logistic limitations. In the vast majority of situations, milk payment to farmers is only based on quantities delivered, although some collection structures try to avoid frauds by testing milk density or acidity (Le Gal et al. 2009). However, these tests do not provide any indication about milk chemical (fats and proteins) or hygienic (level of microbial contamination) quality. This implies that for the majority of smallholder farms there is no direct incentive to improve milk quality, as it won't be rewarded.

The existing references on milk quality in the Maghreb region indicate that fat and protein contents may be acceptable, unless there are conditions of inadequate feeding: in farms relying mainly on concentrates with limited amount of forage, fat contents are generally low (Sraïri et al. 2005). But the most evident characteristics of raw milk in the Maghreb cattle systems is its poor hygienic quality, as its microbial load is generally 100 fold more important than international standards. This is due to poor hygiene at farm level (Aggad et al. 2009; Sraïri et al. 2009c).

As the current context of milk production is characterized by soaring prices of feed and stagnating farm gate milk prices, many farmers argue that milk quality could be a significant issue to improve their margins. Therefore, all the stakeholders within the dairy chains, from farmers to collection co-operatives and dairy processing units are requested to try to tackle that collective challenge. To upgrade the quality of raw milk in the Maghreb region, there is an urgent need to implement transparent milk payment systems. That has to be coupled to regular negotiations between the stakeholders on value chain distribution, and the way to increase it, particularly by promoting milk yield and quality.

This question raises another challenge for the dairy chains in the Maghreb area, which is in relation to the implementation of support programs destined to farmers. In fact, with the huge numbers of non-specialized smallholder cattle farms, the feasibility of on farm follow-up and intervention is quite limited. The traditional State technical services withdrawal from the monitoring of farms has even worsened the situation, like in many other developing countries (Kidd et al. 2000). New channels of farmers learning and innovation are progressively emerging, like the collection co-operatives, but it may take time until they efficiently contribute to assist all the farmers (Faysse et al. 2012). As many shortfalls still characterize dairy farming, from insufficient and imbalanced dietary rations for milking cows to inadequate hygienic practices, on-farm support programs can be practical tools to improve milk yield and increase farmers' income. Therefore, there is an urgent need to place support programs targeting dairy smallholder farms on the forefront of the agricultural policy agenda, if the expected increases in the milk output are effectively sought. Such programs should give a top priority to a continuous assessment of the nutrients' content in the dietary rations used by farmers and try to balance that content with the effective requirements of milking cows. To achieve this aim, and given that both rations and nutrients' requirements of cows are variable throughout the year, a proximity monitoring of farms is necessary (Sraïri et al. 2011).

In fact, all the Maghreb countries have planned a steady increase in their milk output in the near future. For the specific case of Morocco, an overall agricultural strategy, named the "Green Morocco Plan" has been implemented in 2008, and it projects the raw milk output to reach 5 million tons by year 2020 (MAPM 2008). Such a figure implies more than the double of the current milk production. The "Green Morocco Plan" is mainly based on optimistic hypotheses, where important dairy projects based on farms with more than 500 cows may be settled in the most favorable areas of the country, i.e. large scale hydraulic schemes and rain fed regions with more than 600 mm of average rainfall. However, the "Green Morocco Plan" almost does not mention the necessary increase in fodder quantities that would be needed to get such an increase in milk productivity. That constitutes another significant challenge for the dairy chains in the Maghreb countries, as the "Forage Revolution" is quite inexistent and heavy imports of feed being planned to improve animal production (Abdelguerfi and Ameziane, 2011).

Another significant challenge for the Maghreb dairy chains is in relation to the upgrading of the dairy products' offer and consumption. This is linked primarily to the evolution of households' incomes. Currently, the average consumption of dairy products remains limited and mainly dominated by drink milk. If it is to diversify towards more dairy derivatives with higher value, such as cheese, yogurts and light products, incomes will have to be improved, particularly for the vast majority of households with revenues not exceeding 3,000 US $ annually. Moreover, it will also mean that the local dairy processing units may get the necessary know-how to improve their products. That may also be prompted by the ongoing emergence of multinational dairy brands within the Maghreb markets, through joint-ventures with local units. In fact, in Algeria, Morocco and Tunisia, global dairy operators like "Danone", "Nestlé" and "Sodiaal" are beginning to invest and their brand names are becoming common. Such a trend might also assist the overall upgrading of the local dairy chains, given the international standards brought by these operators with regard to milk quality and processing.

Last but not least, the free trade agreements being adopted by the countries of the Maghreb region with the EU and the US may also bear significant consequences for their dairy chains. They may induce the progressive decrease of the high duties levied on the dairy products, particularly in Morocco and Tunisia. That would also mean the end of the protection of the local market, with an acute competition from imported goods. Free trade agreements imply a vast array of measures which need to complement the agricultural reform efforts and preserve the economic and social stability in a sector of the economy which almost employ half of the population.

CONCLUSIONS

The dairy chains in the Maghreb countries represent important players in the global dairy market, particularly because of the leading position of the Algerian imports of milk powder. These dairy chains will continue to grow, as a response to the demographic expansion and the expected improvement of the *per capita* Gross Domestic Product (GDP), which will certainly induce a diversification in the dairy products' consumption towards more high value derivatives (yogurts, cheese and butter). To achieve such a growth, more raw milk will be needed, which implies the implementation of sustainable dairy cattle systems. Therefore, additional efforts will be needed in cattle farms to improve the water productivity through milk and meat, as the limited water resources certainly constitute the most important threat to the increase of milk output. This is particularly true in irrigated schemes located in arid regions, where non sustainable rhythms of use of groundwater have already induced the collapse of dairy farming and its delocalization in areas with more rainfall. On another hand, sound on-farm support programs targeting in priority smallholder units will be urgently needed, as they will surely remain the main actors in dairy farming. Such programs should aim to promote the overall farm performances, from irrigation efficiency, to fodder biomass yield and its conversion to milk and meat, given that such farms are rarely dairy specialized. In addition, more efforts to promote equity and a fair distribution of incomes throughout the dairy chains will be needed. This will have a direct link with the implementation of good governance to try to tackle the collective challenges lying ahead: transparent assessment and remuneration of milk quality, frequent negotiations between the stakeholders (numerous smallholder farmers, milk collection co-operatives and dairy processors) on the sensitive issue of raw milk price, etc. Given the rising volatility of feed and food prices in global markets, the promotion of autonomous dairy chains will be required. That may also be achieved through more regional integration of the national dairy chains between North African countries, by enhanced exchanges of expertise and the encouragement of dairy products' trade. To realize such a goal, a necessary political will be required, to overcome to current cost of "non Maghreb" - almost 5% of GDP growth according to many economy experts of this region -, which already hampers the effective economic growth of the whole region. That might be a necessary condition to favor food security, promote dairy products' consumption and avoid risks of social unrest.

AUTHORS' CONTRIBUTIONS

MTS planned the work that led to the manuscript; MTB and KK participated in the interpretation of the results; MTS, MTB and KK wrote the paper. All authors read and approved the final manuscript.

REFERENCES

1. Abdelguerfi A, Ameziane TE (2011) Interactions between cereal cropping systems and pastoral areas as the basis for sustainable agricultural development in Mediterranean countries. In: Lemaire G, Hodgson J, Chabbi A (eds) Grassland productivity and ecosystem services, UK: CAB International Oxfordshire. pp 261-270
2. Aggad H, Mahouz F, Ahmed Ammar Y, Kihal M (2009) Evaluation de la qualité hygiénique du lait dans l'ouest algérien. Rev Méd Vét 160:590-595
3. Aït Amara H (2009) Quel futur alimentaire pour l'Algérie?. Algiers: Editions Mille Feuilles. p 142
4. Ait El Mekki A, Tyner W (2004) The Moroccan-American free trade area: effects on the agricultural and food sectors in Morocco. Paper presented at the 7th annual conference on global economic analysis. Washington, DC: 'Trade, Poverty and Environment'.
5. Annicchiarico P, Abdellaoui Z, Kelkouli M, Zerargui H (2005) Grain yield, straw yield and economic value of tall and semi-dwarf durum wheat cultivars in Algeria. J Agric Sci 143:57-64
6. Atti N, Rouissi H, Othmane MH (2006) Milk production, milk fatty acid composition and conjugated linoleic acid (CLA) content in dairy ewes raised on feedlot or grazing pasture. Livest Sci 104:121-127
7. Ben Mahmoud A, Jemni N (2008) Supply chain management implementation in Tunisia: an exploratory study of the dairy sector. Int J Log Sys Manag 4:469-485
8. Bencharif A (2001) Stratégie des acteurs de la filière lait en Algérie: état des lieux et problématique. 25-44 http://eco4.voila.net/filierelaitenalg.pdf (Accessed 18.08.2012)

9. Benchelah AC, Maka A (2008) The nutritional value of dates. Phytothérapie 6:117-121

10. Benyoucef MT (2005) Diagnostic systémique de la filière lait en Algérie: organisation et traitement de l'information pour l'analyse des profils de livraison en laiteries et des paramètres de production des élevages. El-Harrach, Alger: Thèse de doctorat d'Etat. Institut National Agronomique. 396 p.

11. Benyoucef MT, Abdelmoutaleb M (2010) Indicateurs de la technicité des éleveurs et canaux de vulgarisation dans des élevages bovins laitiers de la région centre (Algérie). Constantine Algérie: Revue Sciences & Technologie de l'Université Mentouri.

12. Benyoucef MT, Boubekeur A (2010) Approche d'étude de la conduite des elevages laitiers dans l'environnement des oasis de la wilaya d'Adrar. Communication aux 8èmes Journées des sciences vétérinaires Ecole nationale supérieure vétérinaire El-Harrach Alger. 18 et 19 avril 2010. http://www.ensv.dz/spip.php?article22 (Accessed 20.08.2012)

13. Cheriet F, Le Roy F, Rastoin J-L (2008) Strategic alliances between mutinational firms and SMEs: the case of danone - djurdjura in Algeria. Revue Internationale PME 21:35-68

14. CIA (2010) CIA world factbook. https://www.cia.gov/library/publications/the-world-factbook/ (Accessed 17.08.2012).

15. Davis DK (2007) Desert 'wastes' of the Maghreb: desertification narratives in French colonial environmental history of North Africa. Cult Geogr 11:359-387

16. Delgado CL (2003) Rising consumption of meat and milk in developing countries has created a new food revolution. J Nutr 133:3907S-3910S

17. Dhehibi B, Gil JM (2003) Forecasting food demand in Tunisia under alternative pricing policies. Food Policy 28:167-186

18. DRDPA (2011) Direction de la régulation du développement de la production agricole (Ministère de l'agriculture et du développement rural). Algiers, Algeria: Rapport 2010. p 124

19. Dunne J, Evershed RP, Salque M, Cramp L, Bruni S, Ryan S, Biagetti S, Di Lernia S (2012) First dairying in green Saharan Africa in the fifth millennium bc. Nature 486:390-394

20. Errahj M, Kuper M, Faysse N, Djebbara M (2009) Finding a way to legality, local coordination modes and public policies in large scale irrigation schemes in Algeria and Morocco. Irrig Drain 58:S358-S369

21. FAO (Food and Agriculture Organization) (2012) Statistics. http://faostat.fao.org/site/342/default.aspx . (Accessed 20.08.2012)

22. Faysse N, Errahj M, Kuper M, Mahdi M (2010) Learning to voice? The evolving roles of family farmers in the coordination of large scale irrigation schemes in Morocco. Water alternatives 3:48-67

23. Faysse N, Sraïri MT, Errahj M (2012) Local farmers' organisations: a space for peer-to-peer learning? The case of milk collection cooperatives in Morocco. J Agric Educ Ext 18:285-299

24. GIV Lait (Groupement Interprofessionnel des Viandes et du Lait) (2012) Secteur de l'élevage en Tunisie: la filière lait. Tunis. http://www.givlait.com.tn (Accessed 20.08.2012)

25. Haut Commissariat au Plan (Morocco) (2001) National survey on consumption and households' expenditures. Rabat, Morocco. p 404 http://www.hcp.ma/downloads/Niveau-de-vie-et-pauvrete_t11884.html (Accessed 20.08.2012)

26. Henry CM (2004) Algeria's Agonies: Oil rent effects in a bunker state. J North African Stud 9:68-81

27. Jazi S (2003) Confiance et qualité perçue: le cas du secteur laitier marocain. Rev Fr Gest 144:65-81

28. Kadi SA, Djellal F (2009) Feeding autonomy of dairy cattle farms in Tizi-Ouzou area. Algeria: Livestock Research for Rural Development, 21. http://www.lrrd.org/lrrd21/12/kadi21227.htm (Accessed, August 15th, 2012)

29. Kidd AD, Lamers JPA, Ficarelli PP, Hoffman V (2000) Privatising agricultural extension: *caveat emptor*. J Rural Stud 16:95-102

30. Laftouhi NE, Vanclooster M, Jalal M, Witam O, Aboufirassi M, Bahir M, Persoons E (2003) Groundwater nitrate pollution in the Essaouira basin (Morocco). Comptes Rendus - Géoscience 335:307-317

31. Lampietti JA, Michaels S, Magnan N, McCalla AF, Saade M, Khouri N (2011) A strategic framework for improving food security in Arab countries. Food Sec 3(Supplement 1):S7-S22

32. Le Gal P-Y, Kuper M, Moulin C-H, Sraïri MT, Rhouma A (2009) Linking water saving and productivity to agro-food supply chains: a synthesis from two North African cases. Irrig Drain 58(S):320-S333

33. Lionello P, Malanotte P, Boscolo R (eds) (2006) Mediterranean climate variability Amsterdam: Elsevier.

34. Madani T, Mouffok C (2008) Production laitière et performances de reproduction des vaches Montbéliardes en région semi-aride algérienne. Revue d'Élevage et de Médecine vétérinaire des Pays tropicaux 61:97-107

35. Madani T, Hubert B, Vissac B, Casabianca F (2002) Analyse de l'activité d'élevage bovin et transformation des systèmes de production en situation sylvopastorale algérienne. Rev Élev Méd vét Pays trop 55:197-209

36. MAPM (Ministry of Agriculture and Marine Fisheries) (2008) Plan Maroc vert: premières perspectives sur la stratégie agricole. http://www.vulgarisation.net/plan maroc vert.pdf (Accessed 15.8.2012)

37. MAPM (Ministry of Agriculture and Marine Fisheries) (2011) Situation de l'agriculture marocaine n°9. Rabat, Morocco. p 203

38. Mehio Sibai A, Nasreddine L, Mokdad AH, Adra N, Tabet M, Hwalla N (2010) Nutrition transition and cardiovascular disease risk factors in Middle East and North Africa countries: reviewing the evidence. Ann Nutr Metab 57:193-203

39. Office des Changes (Morocco) (2012) Moroccan imports statistics: milk and dairy products, year 2010. http://www.oc.gov.ma/portal/Statistiques/CommerceExterieur/view_EComCHSH.php (Accessed 20.08.2012)

40. Office National des Statistiques (ONS) Algeria (2010) Données statistiques sur la consommation des produits agro-alimentaires en Algérie. http://www.ons.dz

41. Ørskov ER (1999) Supplement strategies for ruminants and management of feeding to maximize utilization of roughages. Prevent Veter Med 38:179-185

42. Patricola C, Cook K (2010) Northern African climate at the end of the twenty first century: an integrated application of regional and global climate models. Clim Dyn 35:193-212

43. Population Reference Bureau (2010) 2010 World population data sheet. Washington: USAID.

44. Rejeb Gharbi F, Lahsoumi R, Gouhis R, Rached Z (2007) Rentabilité économique de l'élevage laitier en Tunisie: cas des gouvernorats de l'Ariana et de Mahdia. Biotechnol Agron Soc Environ 11:211-223

45. Rekhis J, Saaidane F, Laamouri M, Ben Hamida K, Mabrouk WD, Slimane N (2007) Participatory rural appraisal in smallholder dairy systems in Tunisia. Trop Anim Heal Prod 39:619-626

46. Rijsberman FR (2006) Water scarcity: fact or fiction? Agri Water Manag 80:5-22

47. Schilling J, Korbinian PF, Hertig E, Scheffran J (2012) Climate change, vulnerability and adaptation in North Africa, with focus on Morocco. Agric Ecosyst Environ 156:12-26

48. Smith AB (1992) Origins and spread of pastoralism in Africa. Annu Rev Anthropol 21:125-141

49. Sraïri MT (2011) Le développement de l'élevage au Maroc: succès relatifs et dépendance alimentaire. Le Courrier de l'Environ de l'INRA 60:91-101 http://www.inra.fr/dpenv/pdf/C60TaherSrairi.pdf (Accessed 27.09.2012)

50. Sraïri MT, Chohin Kuper A (2007) Conséquences de la libéralisation des marches sur les opérateurs de la filière laitière au Maroc. Revue d'Elevage et de Méd vétér des Pays tropicaux 60:177-187 http://remvt.cirad.fr/CD/derniers_num/2007/EMVT07_177_187.pdf (Accessed 20.08.2012)

51. Sraïri MT, El Khattabi M (2001) Evaluation économique et technique de la production laitière intensive en zone semi aride au Maroc. Cahiers Agri 10:51-55 http://www.jle.com/fr/revues/agro_biotech/agr/e-docs/00/00/EB/07/article.phtml (Accessed 20.08.2012)

52. Sraïri MT, Farit Y (2001) Studies on dairy cattle reproduction performances in Morocco based on analysis of artificial insemination data. Tropicultura 19:151-155

53. Sraïri MT, Ilham A (2001) L'élevage laitier en zone irriguée et sa sensibilité à l'aléa climatique: cas du Maroc. In: Guessous F, Rihani N, Ilham A (eds) Livestock production and climatic uncertainty in the Mediterranean, The Netherlands: Wageningen Pers. pp 321-327

54. Sraïri MT, Karbab A (2010) Consommation de lait et de produits laitiers dans la ville de Rabat (Maroc): effets des facteurs socio-économiques. Tropicultura 28:211-216 http://www.tropicultura.org/text/v28n4/211.pdf (Accessed 21.08.2012)

55. Sraïri MT, Kessab B (1998) Performances et modalités de production laitière dans six étables spécialisées au Maroc. INRA Prod Animales 11:321-326

56. Sraïri MT, Hasni Alaoui I, Hamama A, Faye B (2005) Relations entre pratiques d'élevage et qualité globale du lait de vache en étables suburbaines au Maroc. Revue de Méd Vétér 156:155-162 http://www.revmedvet.com/2005/RVM156_155_162.pdf (Accessed 20.08.2012)

57. Sraïri MT, Rjafallah H, Kuper M, Le Gal P-Y (2009) Water productivity through dual purpose (milk and meat) herds in the Tadla irrigation scheme. Morocco, Irrig Drainage 58:S334-S345

58. Sraïri MT, Kiade N, Lyoubi R, Messad S, Faye B (2009) A comparison of dairy cattle systems in an irrigated perimeter and a suburban region: case study from Morocco. Trop Anim Heal Prod 41:835-843

59. Sraïri MT, Benhouda H, Kuper M, Le Gal P-Y (2009) Effect of cattle management practices on raw milk quality on farms operating in a two-stage dairy chain. Trop Anim Heal Prod 41:259-272

60. Thomas RJ (2008) Opportunities to reduce the vulnerability of dryland farmers in central and west Asia and north Africa to climate change. Agric Ecosyst Environ 126:36-45

61. Tunisian Customs (2012) Tariffs and nomenclatures: milk and dairy products. http://www.douane.gov.tn/index.php?id=667&choix=1&chap=04 (Accessed 21.08.2012)

62. UNDP (2009) Human development report 2009 – overcoming barriers: human mobility and development. New York: Palgrave Macmillan.

63. Valeeva IN, Huirne RBM, Meuwissen MPM, Oude Lansink AGJM (2007) Modeling farm-level strategies for improving food safety in the dairy chain. Agric Syst 94:528-540

64. Vermorel M, Coulon JB (1998) Comparison of the national research council energy system for lactating cows with four European systems. J Dairy Sci 81:846-855 |

65. Wada Y, van Beek LPH, Bierkens MFP (2012) Non sustainable groundwater sustaining irrigation: a global assessment. Water Resour Res 48:

66. World Bank (2012) Population 2010. http://siteresources.worldbank.org/DATASTATISTICS/Resources/POP.pdf (Accessed 22.08.2012)

Citations

CHAPTER 1

Pérez de León AA, Teel PD, Auclair AN, Messenger MT, Guerrero FD, Schuster G and Miller RJ (2012) Integrated strategy for sustainable cattle fever tick eradication in USA is required to mitigate the impact of global change. Front. Physio.3:195. doi: 10.3389/fphys.2012.00195.

CHAPTER 2

Matthew R Dzialak, Chad V Olson, Stephen L Webb, Seth M Harju, and Jeffrey B Winstead, Incorporating Within- And Between-Patch Resource Selection in Identification of Critical Habitat for Brood-Rearing Greater Sage-Grouse, doi:10.1186/s13717-015-0032-2.

CHAPTER 3

Kesang Wangchuk and Jigme Wangdi, Mountain Pastoralism in Transition: Consequences of Legalizing Cordyceps Collection on Yak Farming Practices in Bhutan, doi:10.1186/s13570-015-0025-x.

CHAPTER 4

David Oseguera Montiel, Nícola Maria Keilbach Baer, Akke van der Zijpp, Chizu Sato, and Henk Udo, 'It is Better to Herd than be Herded': Making a Living with Goats in the Bajío Region, Mexico, doi:10.1186/s13570-014-0009-2.

CHAPTER 5

Joshua Sikhu Okonya, Robert OM Mwanga, Katja Syndikus, and Jürgen Kroschel, Insect pests of sweetpotato in Uganda: farmers' perceptions of their importance and control practices, doi:10.1186/2193-1801-3-303.

CHAPTER 6

Laxman Singh Kandari, Vinod Kumar Bisht, Meenakshi Bhardwaj, and Ashok Kumar Thakur, Conservation and Management of Sacred Groves, Myths and Beliefs of Tribal Communities: a Case Study from North-India, doi:10.1186/s40068-014-0016-8.

CHAPTER 7

Anoop Yadav, Renuka Gupta, and Vinod Kumar Garg, Organic Manure Production from Cow Dung and Biogas Plant Slurry by Vermicomposting under Field Conditions, doi:10.1186/2251-7715-2-21.

CHAPTER 8

Mohamed Taher Sraïri, Mohammed Tahar Benyoucef, and Khemais Kraiem, The Dairy Chains in North Africa (Algeria, Morocco and Tunisia): from Self Sufficiency Options to Food Dependency?, doi:10.1186/2193-1801-2-162.

Index

U

V

W

Y

Z